TO COLIN SHAW.

'For old times' sake'.

And the vessel that he made of clay was marred in the hands of the potter: and so he made it again, another vessel.
Jeremiah 18, V. 4

"Surely we have no right to deprive an animal of basic functions which belong to the essential dignity of its species—for example, the ability to lie down or to stretch its body. This is to treat an animal as a thing—to deny it the dignity God gave it. It may be cheaper to produce veal or chicken or eggs by these means. If the result is sub-standard food, it is also sub-standard humanity".

     Hugh Montefiore, Bishop of Kingston,
      in his sermon for St. Francis Day.

Price two pounds or forty shillings.
$6 in U.S.A.

© Copyright in this work is the property of Matthew A. Thompson

Published by MATTHEW A. THOMPSON,
Shipton Poultry Farm, Bridport, Dorset.
Telephone Burton Bradstock 327
ISBN 0 9506199 0 6

Front cover design by Mary G. Thompson
Constructional drawings by David M. Thompson

A copy of this book has been deposited in the Library of Congress, Washington, D.C.

The Organic Poultryman

By

Matthew A. Thompson

FOREWORD

I write this Foreword with the greatest pleasure as I know the reader will enjoy, as I did, every moment of 'The Organic Poultryman'. Whether or not you want to keep poultry or agree with the views expressed I am sure you will find great enjoyment in the forthright way Matthew Thompson expresses himself and will end up with great admiration for the energy and enthusiasm which has carried him forward through a lifetime of experimentation whilst still farming and earning a living.

Matthew Thompson is a rugged individualist, who has a healthy disrespect for orthodoxy, bureaucrats and vast industrial empires. He reminds me very much of the great William Cobbett who in his 'Rural Rides', written nearly two centuries ago, berated the stupid interference of politicians, businessmen and gentry in agriculture. Matthew Thompson follows in that tradition too in being first and foremost a completely practical person who only advocates for others what he has done himself—and usually for a long time at that—and he also delights in looking ahead—much further than most. His method of poultry keeping is a sweet harmony between the birds, the land and man himself. It conserves by endeavouring to improve every aspect, and never debases one part in a futile attempt to enrich another. There is no doubt that many processes in modern agriculture are wasteful but Matthew Thompson's organic poultry keeping is certainly not. He shows that breeding for improvement can be an unsophisticated procedure and the attainment of good health he rates exceedingly highly—a policy in which I am in complete agreement since we know that a healthy animal of any species always produces most economically, and gives us the best quality of food. He attaches great importance to longevity—a very wise policy—and looks for birds that can be kept for several years, despising that in-built obsolescence that has engulfed so much of life, agriculture and industry.

I do not want to over emphasize the fact that you will enjoy reading the book, whatever are your agricultural intentions, because there are very many points made by Matthew Thompson that should cause the reader to reflect very deeply, both in depth and in breadth.

We must think not just in terms of satisfying the nutritional needs of the advanced nations but of doing it world-wide; it is the very poor countries which urgently need to have more and better poultry and it is they who could get such enormous benefit from Matthew Thompson's practical approach. I know from first hand observation in many 'third-world' countries that many of our highly intensive systems are introduced with disastrous consequences. Though none has achieved the notoriety of the Gambia egg scheme I suspect many have lost just as much or perhaps even more! The primary stock must be hardy under all sorts of conditions and they must be capable of being looked after in a straightforward manner by people who may have a minimum of education or experience. Feeding should be on simple lines and foraging will be of great assistance. The greater the trend to self-sufficiency the more practical and useful will the system be and here that approach will be found.

In life in general, and in the agricultural field especially in poultry keeping, we are persuaded to believe in stardardisation, conformation and belief in high technology. We are warned against believing in anything other than the latest research findings. And as Matthew Thompson suggests, we are easily taken in by commercially interested persuaders. It is fortunate that the brain-washing does not deceive everyone and here the reader will find the absolute antithesis to all those things I have listed and the presentation of an absorbing account of one man's wisdom, experience and practical approach towards poultry keeping the organic way.

Cambridge,

March, 1978.

David Sainsbury

CHAPTER ONE

"COME INSIDE AND LET'S TALK IT OVER"

For many years now my wife and I have shared this house with, among other things, a very large and corpulent spider that goes by, even if it does not answer to, the name of Harold.

At least we have always given him the benefit of any doubt which may exist and assumed him to be the same, identical fellow and not merely one of a long succession of such spiders, all looking alike. If the latter should turn out to be the case, at least we are no differently situated than all those simple folk who for ages got along quite happily while convinced that the prophet Isaiah was one and indivisible. Then the theologians, not content to leave well alone, discovered—or invented—a further half dozen or so; and the faithful have been unsettled and assailed by doubts ever since.

Anyway, there our Harold sits, enthroned amid his web, awaiting the arrival of one stupid or short-sighted fly after another, that he may suck them dry.

I sometimes feel like that spider. For I seem to have a fatal attraction for students, drop-outs from the rat race and all manner of humans each of whom, however much they may differ in other respects, seems to have one aim in life. Every one of them thirsts after knowledge—or further knowledge—about the organic or self-sufficient way of life and in particular where and how fowls come into it or may be happily and successfully incorporated therein.

Not, mind you, that I ever attempt to suck any of them dry. Rather it is the other way round. They try to suck me dry, or at least to pick my brains. Not, again, that I mind. Very few of them ever purchase anything or contribute in any way to my finances. But I am content to share the rich and varied harvest of the years and the voyagings—of my successes and failures, my triumphs and disappointments or near-disasters—with them. And if they do take up a little of my time, well I can always make it up afterwards.

At least that was the situation. But just lately things have got to such a pitch that both my helpers and myself are always desperately short of the commodity just mentioned. And somebody, knowing me to be a veteran broadcaster, scripwriter and journalist, has suggested that it would be a nice idea if, instead of having to trot out pretty much the same things, over and over again, even if to a succession of fresh faces, I were to put all this information and guidance, the riches of experience, into some sort of order and confine it within the covers of a book.

So far so good. We have opened up the dialogue. We have established our respective roles: you as the supplicant, the seeker after help and, if possible, wisdom and inspiration; myself as guide and mentor. In all these situations and relationships I do my best to infect the other person with my own enthusiasm, to fire whatever is latent within them, just waiting to be kindled.

As like as not you find yourself, as the encounter takes place, in a sort of limbo, having left one existence or way of life—or apology for a way of life—behind and not quite taken up another. It may be, also, that you are not altogether clear in your own mind about a number of things. It goes without saying that you are perfectly well aware of what you have come from. And it is because you have found it distasteful, or vapid, or cramping or in one way or another intolerable that all this has come about.

If it is merely a case of escaping from an encompassing state of affairs that one cannot stand a day longer, then the possibilities are endless, though not all of them equally attractive. For instance, in an issue of *The Times* which goes a long way back now, some imaginative and enterprising character was advertising a hole 200ft. deep in some remote European fastness. What the vendor, no doubt, had in mind was that someone who had had enough of trying to make sense of a crazy world and to cope with life as, up to now, he had found it, would decide to take up permanent residence there, having previously booby-trapped or barred the entrance against possible intruders or gatecrashers. Such a one would need to get in quite a stack of provisions and gear of various kinds; and the nature and extent of these would be determined by his own assessment of life expectancy.

It does not appear likely that you who have just taken up this book will feel tempted to go quite as far as that. But it is to be assumed that you now look for more from life than that which you have hitherto enjoyed —or found it impossible to enjoy.

However such a resolve is not in itself sufficient to provide you with a blueprint for mere survival, let alone with the wherewithal to sustain body, mind and spirit over a period of years. On two widely-separated and memorable occasions, turning points in a long and varied career, the bank manager who happened to be in favour at the time has pointed out to me, employing identical words: "It isn't sufficient to have something to live for—you must also have something to live on". It may be unfortunate; some of you may even regard it as brutal; it is nevertheless one of the facts of life. But this does not mean that we need ever, any of us, compromise or descend to that which is unworthy. It is a long time now since I reached the stage at which I was able to say with conviction that it is perfectly possible to eat and at the same time be free.

Let me enlarge upon this. When I first took over this place here in Dorset, many years ago now, I determined that, come what might, I would not compromise with what I knew to be wrong, stupid or unnecessary. From the very outset I was resolved that the clean and moral thing, the thing that made for health, wholesomeness and a proper, all round balance: that would be the thing for me. And if this approach proved to be false or this hypothesis groundless; if, that is, it failed to pay dividends, then so be it—I was prepared to go down with the ship.

As it is I have been vindicated up to the hilt. Financial gain for myself has always come last on the list. But, after all these strenuous years of doing things my own way and listening to nobody, the results are becoming clearer and more gratifying with each day that passes, and more and more folk are coming forward—and welcome they are— to share in the fruits of all this work, occupying innumerable hours.

The other year a certain county poultry advisory officer (a far too grand title for one who has missed his way in life and opted with his eyes open for the softer job of leading people up the garden path instead of the often harder but more satisfying one of setting them right), telephoned me one day. This was the only way in which he could make contact with me, since he knew that he would never be permitted to set foot on the place; nor has he to this day. Anyway, he had just finished reading an article of mine that had been widely and often heatedly discussed in many countries. And he said: "I agree with every one of your conclusions". (He then went on to detail them). "But what can I do? I must go along with the commercial men".

I find such an attitude not only deplorable but utterly incomprehensible. I have never been able to subscribe to the view that there can ever be a right reason for doing a wrong thing. If you know the thing to be right, then you simply go ahead with it at all costs and regardless of the consequences. If, on the other hand, you are satisfied it is wrong, you refuse to have anything to do with it. That is all there is to it.

There are no grounds whatsoever for the assertion that, "Of course we must have factory farming, because there are so many mouths to feed". I heard someone who describes himself as a naturalist utter these very words one afternoon in a children's programme on B.B.C. TV. And they and similar sentiments are frequently aired by those who ought to know better and have no particular commercial axe to grind.

Those who pursue monoculture or employ poison sprays and chemical fertilisers, or who confine and feed hens or calves or whatever in grossly unnatural ways just couldn't care less about all the empty bellies there are waiting around to be filled. And in fact the thing that scares them most of all is, not shortages, but any prospect of a glut. There is, in any case, something fundamentally amiss with a society that, plastic and tasteless chicken and wishy-washy eggs notwithstanding, is prepared to accept as normal or unavoidable such a state of affairs in its midst.

Be all this as it may, you are already committed to a life on the land and have given an earnest of your intentions and sincerity by burning your boats, or some of them anyway. There is no going back or doing anything that might lay you open to a charge of cowardice or timidity. Having ascertained all this, and got to where we have in the book, the natural thing might seem to be for me, the author, to turn my long and wide experience to account by sketching out details of the sort of farm or holding the beginner would do well to set his or her sights on. Alas, most of those who come to me upon such errands are either committed already, having, for better or worse or richer or poorer, already bought, inherited or had a property wished upon them; or else they are severely limited in their choice by the amount of cash they have or can lay their hands on or the availability of suitable sites. The late H. R. Hunter of Winchester, replying to a letter sent him by a woman who required a breeding hen and used up quite an expanse of paper explaining what it had to be like and the many and various conditions it must fulfil, and having no more time to waste than the rest of us, simply replied: 'Madam, that bird is in heaven'. Dream farms are not in the habit of dropping out of the sky every ten minutes.

In a word, they find themselves having to make do with what they are already landed with or are able to come by at reasonably short notice and before the effects of repeated frustrations and rebuffs have had a chance to set in.

It is, of course, true that, given certain basic comforts and conveniences, different places are more or less suitable for different needs and purposes. In my own case, for example, as a breeder, the nearer the local station is the better. Unfortunately it is nowhere near as handy as it used to be. Or, if that sounds too Irish, it is a different station. At one time we could, at a pinch, nip into the car and be there inside ten minutes. But not any more. We are involved, each time, in a journey of twelve miles both ways.

You, dear reader, as they used to say, are hardly likely to be troubled by such considerations. No tyro, surely, is going to be sufficiently naïve to hope or expect to set up as a breeder, on whatever level, just like that.

An hour or two after those words had been put down an exceedingly elderly gentleman dropped in without prior warning to purchase a stock cockerel. Although now resident in Devon, he hails originally from Lancashire and, like so many of us, has in the past spent many hours trapnesting, recording and carrying out those myriad other routine tasks which have traditionally been associated with pedigree work. He, therefore, has at least some inkling of what he is talking about when he gets on to the subject of this book. He also remembers the poultry 'greats' of the past: Tom Barron; Richard Rodwell; Y. Watanabe and the rest. And as he sat in this room he was kind enough to say that not one of them, or the lot in combination, could have achieved what I had.

It is years since he came here and took away some pullets, and he says they still lay every day. Their only fault, in fact, is that they are such good and adventurous foragers, it is necessary, very often, to go to the other end of the farm to look for them and bring them back.

It is a good and satisfying thing to have thus got to the top of the ladder; but such things do not happen overnight. You may succeed in emulating this, and if you do I shall be the first to take my hat off to you—or I would if I wore one. Only it won't be this week, or even next.

With your feet set firmly on the ground, then—and never losing sight of the fact that it is not always spring or summer and that gum boots are often the order of the day—let us see if we can manage to lead you a little further along the path you have chosen.

I have written elsewhere and repeatedly about the evils that are inherent in specialisation and the tendency, which has been with us ever since big business first moved into the barnyard, for far too many birds and far too much equipment, housing and one thing and another to be concentrated in too few hands and places. It was my fellow broadcaster and interpreter of the farming scene to the townsman, the late A. G. Street, who gave the right answer to those who kept on clamouring for agricultural workers to be paid more. This, they claimed, would not only stop the drift from the land, but reverse it. Street pointed out that, far from this being the case, whenever farm wages were raised by government edict, the farmers' only response was to turn away more men than before, cutting their wage bills and installing more machinery and other labour-saving devices.

Many more families could pursue a happy, fulfilled and useful life and career on the land than at present. And the benefits and repercussions of this would go on being felt further and further out, as when a stone is dropped into the middle of a pond. It makes neither economic, social nor any other kind of sense for great, ugly broiler, battery or hatchery complexes to stand amid stretches of empty, henless acres.

To hope to put an end to the deporable state of affairs of which all this is no more than a single facet; to entertain the notion that any kind of a sane, healthy and balanced farming pattern can be brought about through legislation is to indulge in an impractical dream. The growing concern shown in this country, the U.S.A., and all the way from Italy to Sweden, about such matters as rabid intensivism and the quality of produce, and the increasing resistance being put up by housewives all over the place: in these and more of the right sort of education lies our best and perhaps only hope.

It is certainly true that the handful of enormous concerns which have, for several decades now, held such a grip on the market and poultry industry can only continue to wax fat so long as a sufficient number of ordinary folk can be brainwashed into supporting them. And as long as they remain in being, just so long will things stay out of kilter, with great quantities of effluent and manure—much of it, admittedly, of questionable value because of what the stock producing it has been fed on or doctored with—being produced where there is no outlet for it so that, more often than not, it has to be dumped, or flushed down drains or into rivers—and all those henless acres staring us reproachfully in the face.

The late Ralph Wightman rang me up, once, to discuss this problem. But the most constructive offering he seemed able to make was that, while an infernal nuisance, it seemed to be at present unavoidable.

Time was when many of our fields enjoyed, and benefited enormously from, the scarifying effects of the activities of flocks of busy fowls. And their droppings were far better distributed. As the stock was kept in very much smaller concentrations than is the case nowadays there was far less risk of disease or of any of the other ills that attend the kind of husbandry—for the want of a better or worse word—modern technology and the application of highly paid if addled brains has made possible.

In many cases the poultryman or general farmer had only a few birds on the place and so was able to observe them more closely than is the case with those to whom a fowl is no more than a statistic or an entry in the right or wrong side of a ledger. Often he loved his stock and was conversant with them in the true meaning of the expression. And, as I have pointed out repeatedly, love is not only an essential ingredient of the make-up of the real poultryman; it is by far the most important.

Such a one was my mother's father who, surrounded by neighbours that were mainly cottagers and sheep farmers and regarded hens in much the same light as they did the weather—either they laid or else they didn't; and if they didn't there was little anyone could do about it—attracted much attention to himself. When they found out about my grandfather's success with his White Wyandottes, bred from stock originally obtained from Tom Barron of Preston, they came to look upon him as a species of wizard or oracle and would make frequent pilgrimages—in fact some people came from a surprising distance, I remember—to sit in the chimney corner and listen to him and afterwards to go out and just stand and watch these remarkable birds literally queuing up to lay, and in the depths of winter, too.

Neither he nor any of his contemporaries, cast in similar mould, experienced any difficulty in finding customers for the cockerels which their broodies obligingly brought forth along with the pullets required to replenish the laying houses and fill the egg basket. Those not required for, or not suitable for the breeding pen, having been prepared by the industrious and thrifty womenfolk, were collected at the door or delivered, in a basket covered with a snow-white cloth, to houses within the more or less immediate neighbourhood. Or else they were sold from a personal stall, along with butter, eggs and other produce, in the nearest market town. Contrast this with the contemporary scene on the poultry front both here and in America. On both sides of the Atlantic, with every year that passes, millions of newly-hatched males, sexed beforehand where necessary, are gassed and fed to reptiles and suchlike in zoos and places like that. Such material is known in the trade as 'incubator waste'.

This shining example of folly and downright wickedness is presided over, at least on this side, by the National Farmers Union—N.F.U. for short. As this is being written it is seldom possible, even in rural areas, to obtain milk that tastes like milk and not washing up water and has not been subjected to pasteurisation and generally messed about with. Indeed we are busily engaged in raising a generation of folk who will never be able to remember what the genuine article tastes like, as they have never had an opportunity to find out. And at this very moment innumerable housewives are busy spreading on bread out of which much of the goodness has been extracted, and that nevertheless costs only slightly less than gold bricks once did, butter from Germany, Denmark and God knows where else, with a herd of cows only a few yards down the road. Only in a crazy, greedy and materialistic world could such things happen.

I give those to whom these words are being addressed credit for sufficient intelligence and enough feeling for what is right and wrong, for what belongs and what does not, to want no truck with any such practices, or with high cost systems one of the more bizarre outcomes of which is that, even as they sit and read this, those thugs who run **things in the Soviet Union are licking their chops and crying into their** samovars—tears, not of sorrow but of joy mingled with astonishment at the actions of a political set up disguised as an economic one that can almost literally give away shiploads of produce in about the rummest game of snakes and ladders ever invented.

Some of you, to be sure, may be more interested in and have a greater aptitude for the table side, or may even be situated, as for instance, a few of my customers in East Sussex and elsewhere are, in localities where the demand for genuine outdoor chicken is insatiable. In that case your role will be the reverse of my late grandfather's, and it will simply be a case of disposing of the pullets at such an age as this can profitably and conveniently be done without allowing the whole enterprise to get lop-sided.

And perhaps this is as good a place as any at which to clear the way forward with a brief discussion as to what we both understand by such expressions as 'self-sufficiency'. I think we are all of us pretty well agreed upon what we mean when we employ the word 'organic' within the terms of reference of a book such as this.

The trouble is that so many of these things, beneficial, sound and sensible in themselves and as originally conceived, tend to degenerate into cults or what our American friends call fads. There is never any shortage of those around who are only waiting to clamber on the latest band-wagon if they can see any kudos or advantages to be gained thereby. Thus the tenets of the late Sir George Stapledon, the father of modern ecology who carried out in person some of his grassland experiments on our cow pasture soon after my brother had taken over the farm from me, were initially received by most of the world in something approaching stony silence. And his book remained neglected for years and until revived, fairly recently, by Charles Knight. But those who, like myself, have survived him and have been plugging away at this come rain, shine, indifference or hostility, from the very beginning, saw it become respectable and even fashionable, so that it was the thing to believe in—or at least make out you believed in—if you wanted to ingratiate yourself with the right people.

Self-sufficiency is the latest on the list. The B.B.C. TV people have even seized the opportunity to use it as a springboard for a very light and bubbly soap opera. I trust nobody expects to learn anything from this series, except that Penelope Keith, the archetypal rich-bitch, is a very talented and versatile actress. It is patently obvious to any save the most dim-witted or successfully brainwashed that this pair of woolly-headed, pathological optimists could not possibly make ends meet and pay even the minimum of household bills they are supposed to run up out of the proceeds from whatever they are able to raise on those few square yards of garden in Surbiton.

Admittedly it is infinitely more harmless than that other and much longer-running soap opera, *The Archers*. Its worst and most dangerous and insidious feature consists, of course, of the official plugs which are as frequent and as cleverly-engineered as the propaganda put out by the late Dr. Goebbels who, along with his master, believed that, if only one kept on repeating the same thing, not as an exhortation but as if it were part of the natural order of things, the populace would in time come to assimilate it along with its meat and drink. And all this without having uttered a word about such things as the spurious and unconvincing antics, observations and dialect of Mr. Walter Gabriel.

I hope to goodness no one concerned with the organic movement entertains any notion of learning how to farm, or even how to live, from it. It certainly contains a plethora of object lessons in what not to do or have any truck with, if these can be recognised for what they are.

It can be stated quite categorically and without the slightest fear of rebuttal that no good has ever come of official interference in agriculture. On the contrary an incalculable amount of harm has resulted. For my own part, for officialdom and bureaucracy, in whatever shape or form, guise or disguise, these may be encountered, I have nothing but contempt; nor, as many of those concerned would be able to testify—though this is the very last thing they would want to do—have I ever made any attempt to hide it.

Of course only a small proportion of those in this country—or, for that matter, in America, calling themselves farmers may, within any sane or traditional frame of reference, be properly so described. And it follows that exactly the same applies to farms. Having taken care to make which distinction between real farmers and non-or-pseudo-farmers, let it be stressed that the urgent, the desperate, life-or-death need is for those who till or raise stock on the land, on whatever scale, to break out of the prison many of them have only themselves to blame for. Anyone concerned in any way with the soil needs his head seen to if he has not by this time woken up to the fact that the only long-term result which can follow a visit, cap in hand, to Whitehall is more and more chains. It was the late A. G. Street who, some years back when contingents of West Country farmers—or alleged farmers—converged upon Whitehall one day to demonstrate in a most undignified manner and to demand 'the rate for the job', was heard to observe, trenchantly: "The bloody fools! They ought to bide at home and get on with the job of farming, instead of trying to get theirselves mixed up in politics".

The sturdy, independent grandfathers of many of these misguided folk would be ashamed and disgusted if they were able to come back and see them thus behaving like beggars. The British farmer has no future unless, without too much hesitation or delay, he cuts loose. He must lose no time in learning to stand on his own feet, eschewing all sops, subsidies, 'support payments', 'guaranteed prices' and the rest as he would the devil, and sending every official, major as well as minor, who has the temerity or foolhardiness to show his face on the place away with a flea in his ear.

This book holds no brief on behalf of those who would have us believe that agriculture ought to be an adjunct of government to a greater extent than it is at present, or at all. It is, therefore, a complete waste of time for anyone to look within these pages for any information as to what, for instance, their 'statutory obligations' are or any guidance on how to go about meeting them.

Quite aside from all considerations of financial gain and so on, the more odd and out of the way corners of the land copies of this book succeed in penetrating the better I shall be pleased. Only in such ways can a counterblast be provided to all the brainwash and propaganda that flows out, like the contents of a broken sewer, from the host of magazines and other publications whose outsides are got up so as to kid those whose eyes have not yet been opened that they deal with something worthy to be described as farming, but whose insides tell a different story to the men and women who are able to see things as they really are, and from those who, as we have already observed, have found it expedient to move into the barnyard.

Just one, final word on the subject of self-sufficiency. My wife believes it ought to mean being entirely dependent for a livelihood upon what is produced in one way or another within the immediate bailiwick. Admittedly scale enters into it and a lot depends on acreage or the total capital investment. But the great majority of those we know who have most to say about self-sufficiency are in much the same position as those American youngsters who can afford to drop right out of society always so long as the allowances continue to arrive regularly from the folks back home, who in turn are only enabled to keep this up because they have not yet seen the light.

We know a very large number of couples who have, within the last year or two, in one way or another come by a few acres of English soil, and without exception they appear to fall into one or another of a certain group of categories. Either they have retired and are therefore in a position to get by whether or not the hens lay and the pigs wax fat and the lettuces and what not flourish or even if the whole lot of them sicken or wither away and die, or else they have an army pension to depend on, or they teach or lecture or write books and make TV programmes or in one way or another are largely or wholly independent of the efforts and good intentions of the hens and things.

In the beginning. . . . . . .

The author's wife takes a hand in the proceedings, and begins to make an impression

A lesson from the schoolmistress. A budding gardener takes his first steps under the watchful eye of the author's wife.

Open Day, 1962. This was the first experiment ever carried out to determine whether Russian comfrey, rich in potash, could be successfully used in connection with early potatoes. The author doubled his crop. Variety: Arran Pilot.

CHAPTER TWO

THE RAW MATERIALS. OR, HORSES FOR COURSES

The casual manner in which at least a proportion of the poultry keepers or would-be poultry keepers whom, in the course of the daily round, I encounter, or hear of or from, or whose burnt fingers are held up for my inspection, approach the business of taking in new stock is a never-failing source of astonishment to me.

In fact some folk give less thought to, and go to even less trouble over, the choosing of their future layers or breeders than countless humans do over the all-important matter of the choosing of a life partner.

Of course plenty of enthusiasts come here, or telephone or write, who are perfectly well aware that, just as a bad cow eats as much as a good cow, if not more, so all the various charges go on, week in and week out; and at least as much labour has to be put in, and time spent, over a field or a house full of passengers as over a flock every last member of which is pulling her weight. So far as breeding is concerned, although I may tend to give a different impression in a later chapter devoted to this subject, in the ordinary way, if you want sound foundation stock, whether it be cattle, horses or any other class of livestock that is involved or being considered, you have to pay for it. You can only hope to breed one sort of thing from rubbish, and that is more rubbish.

Nevertheless this doorstep upon which I am at present sitting to scribble this down preparatory to typing it out has in the past been the scene of quite a few verbal tussles. And the telephone at my back has also, and on even more occasions, provided a link between myself and those who thought my prices too high. "Oh, I can pick up a Rhode Island Red stock cockerel for less than that". Or, "I know where I can lay my hands on any number of point of lay pullets at thirty bob apiece". Another variation takes the form of: "But I only want it for such and such a purpose".

To all such I have but one answer to give: "It's your own money you are proposing to lay out, or speculate, or chuck away as the case may be. If you are indeed able to obtain comparable stock, as long-lived and productive as that which I am offering, at a lower cost, then jolly good luck to you".

But don't, for goodness sake, ever lose sight of the fact that these birds and yourself are going to be together for a long time. At least, as in the case of husband and wife, that is usually the idea. And, as with matrimony, either you are destined to spend a great many hours hating the sight of one another and secretly or openly wishing the person in Hell who first introduced you, or else you are going to enjoy and profit from each the other's company immensely, and get along swimmingly.

What, then, to choose—always assuming, that is, that there's a wide enough range available to admit of any particular choice—and what to avoid: that is the question. And it is to this that we must now address ourselves.

The late George Henderson, author of that classic, *The Farming Ladder*, observed that no one could be said to have succeeded at anything connected with the land and livestock until he had taught it to others and got the message over. This may sound a trifle enigmatic. Yet it is a fact that the very skills and features we old hands are most prone to take for granted, so that for someone like myself to know the difference, for example, between a good horse or cow or stock cockerel and a poor or middling one becomes second nature, are those we should remain conscious of if we are to become successful communicators.

The task of getting those who have a hankering to follow in our footsteps to see what we see and feel what we feel seems at times insuperable. But it has, somehow, to be attempted. Let us begin, then, with a consideration of the simplest and most obvious things to look for or to steer clear of.

Years ago I drove one morning to the outskirts of a famous northern market town some 24 miles away from where I was then living and farming. The occasion was a closing down and dispersal sale. The vendor, a youngish man, had been running the place as a poultry farm for the past few years, but had finally run into heavy weather and had to get out.

Being, for all my own youthfulness, of a suspicious turn of mind, and having a feeling that there was more to all this than met the eye or appeared in the catalogue, I turned up early so as to have an opportunity of making a tour of the outlying fields. None of the many others who attended the sale had thought of doing this, and so I had it all to myself and there were none to inquire into my motives. But I was glad I had done so, because it provided a clue to what had contributed towards the bringing about of the forced sale.

Every hedge back, each hollow or clump of bushes, held its quota of dead birds. Some had been flung there only recently; others were in various stages of decomposition. The stock which the disillusioned possessor of all these deceased hens and pullets had penned up ready to be auctioned off were those who had not yet succumbed to whatever disease it was that was rampant or were not going to and, for the present at least, appeared in reasonably good shape.

This kind of thing has its parallel, on a different level, in the show ring, where you will see animals or birds on display that may look splendid but have been carefully selected often from a very much larger number and are not therefore representative of that breeder's general stock or the family to which they belong. From a breeding point of view such practices amount to a form of deception.

I never tire of saying to people who come here for advice: "For goodness' sake, if you want to find out what sort of stuff a particular breeder turns out, don't go to see him. By one means or another get hold of the names and addresses of a few of his customers. Call on them and have a look at the growers in the fields and the pullets in the laying houses—in other words, the material he has supplied within a work and production environment and doing—or failing to do—the job for which they were bought and sold. Ask them all the questions you can think of.

The breeder can show you anything, or hide anything, or tell you anything".

Not many weeks before this was written an order came to hand, couched in formal terms, from a man whom I had previously never heard of, farming in a different part of the country. All the information he vouchsafed was that he had received some details of my stock from the Soil Association, and maybe one of my Price Lists. Anyway, his needs were dealt with promptly and in the same way as anyone else's would have been. Only afterwards did I receive a very nice letter from him, which this time did not begin with 'Dear sir', in which he thanked me for the prompt and efficient way I had dealt with his order and said how pleased he was with the chicks. Then he went on to reveal that he was Chairman of an organic body covering a large territory and would have no hesitation in recommending me and my stock to any of his members who might in the future be in the market for such things, etc., etc.

Don't be misled by cleverly-worded adverts or glossy brochures. Few cloak-and-dagger merchants or confidence tricksters in history have been able to match the devilish cunning of those who, from some time, I suppose, around 1930, when poultry keeping ceased to be looked upon as, in the main and with a few exceptions, nothing more serious or important than a source of pin money for the wives and daughters of general farmers and a hobby or modest livelihood for an odd—perhaps in more ways than one—male here and there, and could be seen to bear the marks of an incipient industry, have regarded it as their oyster.

I think it was in 1929—I have lost track and the present Principal, once my next door neighbour, is unable to help—that two hard-working and enterprising young farmers, H. and E. Robinson of Crathorne or Kirklevington near Yarm-On-Tees, a locality already renowned for stock-breeding, won the Harper Adams Laying Test. They happened to be old friends of my father's and I was acting as assistant to them. Our methods seemed revolutionary at the time. Anyway it was about then that the rot first set in and calculatedly misleading adverts began to appear in those publications that were taken in and also read from cover to cover, in the intervals between being sat on as they hid tidily under cushions and things, by the greatest number of likely victims.

Since then the whole process has snowballed. It took an even greater leap forward when Biddy finally became 'emancipated', when, that is, the humble hen, in the great majority of cases, throwing aside all pretence at sticking to her traditional role within the general scheme of things, took her place in the power game, became the pawn of big business and began to figure in take-over bids and to influence the Dow Jones or F.T. Index. And, don't forget, those who are behind the racket now have at their disposal, not only enormously larger funds, but also far more advanced and sophisticated eye, pound and dollar-catching equipment of one sort and another.

No prizes are being offered for correctly guessing who it is that foots the bill for all this. As with the lawyers, a great many of whom have for years been making a real killing and none of whom are anywhere near so innocent or so burdened with moral scruples and such as you and I, these characters are only able to continue to operate, and to holiday expensively in the sun, while the supply of mugs keeps up.

I want this book to go out as a clarion call to all those who have it within their collective power to bring all this wickedness and this blatant and unashamed exploitation to an end. The remedy lies in your own hands, and all that is called for is more and better-directed effort and a greater identity of purpose.

Two of the most useful and profitable arrows in the enemy's quiver are labelled, respectively, VACCINATION and BLOOD-TESTING. These have been a real godsend to large numbers of slick operators. Over the past several decades, millions of day old chicks, few of which were worth house room, have been unloaded on the strength of the blood test alone. The idea intended to be conveyed appears to be that this confers some sort of magical property upon them. And it is reminiscent of the cigarette adverts one observes splashed about all over the place:—

'EVERY PACKET CARRIES A GOVERNMENT WARNING'

It represents an attempt on the part of the huge tobacco concerns to convert a necessity into a virtue. It is not being here suggested that there should be any kind of direct or roundabout official embargo on the smoking of cigarettes, or for that matter the doing of anything. The firms are, nevertheless, landed with this thing, and since they have to print these warnings their way round it is to pretend that they are giving something away, much as in the days of the original, much coveted cigarette cards.

The whole situation, as it affects this particular variety of humbug, can be set out in a few sentences. The object of carrying out a blood test is to make sure one's adult stock is not playing host to what is known as pullorum. This manifests itself in young chicks as B.W.D. or baccillary white diarrhoea. Any breeder who has so little confidence in himself or his stock that he feels the need to go through these particular motions, let alone advertise the fact, ought to lose no time in selling out and looking for some job that is more in keeping with his qualifications and general level of intelligence. He has much in common with the fellow who, before setting out in the morning, informs all and sundry that he can guarantee a fine day, but takes both a raincoat and an umbrella with him, just in case, or the cleric who, from the vantage point of his pulpit on Sunday, assures his flock that no harm can come to them or theirs if they will but put their trust in the All Knowing and All Caring, and hastens, come Monday morning, to obtain the best possible insurance cover for his every last chick and child, good and chattel, stick and stone.

As for vaccination, the reason for the carrying out of which is supposed to be to protect the feathered recipient against such visitations as fowl pest and fowl plague, this amounts to nothing more nor less than a futile attempt to have one's cake and eat it; to go on sinning and abusing nature while escaping the consequences which normally follow upon such a breach and such practices. It is to take a short cut through a one-way street or a blind alley. One day, a few years back now, a photographer called at this address, having come from an address in Fleet

Street. "There's a lot of discussion going on in Fleet Street", he intimated, "about what you wrote on the subject of the vaccination of fowls". "Tell me more", I invited. "And here's a mug of authentic cider to swig at whilst doing so". So he weighed in and informed me that what they could not make out was why I should object to vaccination when it had demonstrably kept alive thousands or millions of birds that would otherwise have succumbed to mareks disease or whatever.

A lesser mortal, or a more impatient or intolerant one, might have upturned the whole flagon and baptised him with it, or shown him the door. However, as already indicated, representatives of just about the whole human spectrum pass through these portals in the course of a year: counts; dukes; baronets; members of Parliament; millers; professors; Drs. of this and that; lecturers; lorry drivers—I will not go on. And I have to accommodate myself to them, or them to me. And so I merely contented myself with pointing out that any disease or condition you have a mind to name can be cured perfectly simply and quickly if only the methods employed are sufficiently drastic. Whether or not whatever you have left at the end of it—always assuming it to have been any good to start with—is worth keeping is another matter altogether. A child can be cured of whooping cough within the space of a few minutes by letting it fall into a swift and deep mill race and walking away. And some writer of doggerel has put into the mouth of the official who was the more immediate instrument of Charles I's rather sudden demise the words:

"I will cure your hacking cough
When I chop your headpiece off".

On this whole question of disease and the maintenance of the health and well-being of one's flocks, readers of the farming press in different parts of the world are already familiar with my own methods. However I am not quite so coy as to persuade myself that there is no one left around who is still in the dark over them and would not benefit from a spot of enlightenment.

For a great many years now I have kept a small nucleus of birds in the neighbourhood of the house. Under the prevailing dispensation, those that have made the grade are allowed to pass from there to a farm we have of over 100 acres from which the eggs are taken to fill the incubators supplying customers' requirements. I have, for the past couple of years or so, had an understudy who, with his wife, looks after all these birds out in the fields and sees to the day to day work under my supervision. Of course most of the spade work had already been done, and the vastly improved strains and breeds made, before he came into it. Just the same his contribution has been of great value and indispensable, and will be more so as times goes on. I therefore welcome this opportunity of acknowledging it.

He has attended a public school, followed by a course at an agricultural college. And it stands to his everlasting credit that he has managed successfully to overcome both these handicaps.

This small nucleus of stock of which we spoke just now has been kept on the same piece of ground year in and year out. For a long time, in fact, the adults lived in the open, without any overhead shelter provided at all, day and night, summer and winter. They either roosted in trees or on a series of rails, each supported by two low posts. The eggs were deposited behind sheets of heavy corrugated iron, leaned up against a wall with straw laid behind them.

Some of the best results I have ever obtained were from pullets kept out in the fields, with no shelter at all except for trees, stone walls and hedges.

But to return to these birds which have their being not many yards distant from where we ourselves live and where the office and headquarters of the whole organisation are. Those from whose eyes the scales have not yet fallen tend to look upon it as a topsy-turvy way of going about things. For the object of the exercise is and always has been to extend an open and standing invitation to any alien army that can manage to effect an entry. Or, to change the metaphor slightly, we write WELCOME on the mat in big, bold letters, so that any bugs of any kind that may chance to be in the offing may enter freely. From the very start of the enterprise here in Dorset it has been my policy to do all that lay within my power to induce worms and intestinal parasites, coccidiosis, both caecal and duodenal, and the bugs of every known and unknown disease and condition. Short, that is, of actually importing stock already infected. In fact my flocks have for years been closed, so there is no danger of this. I would not, in any case, know from where to obtain stock that would improve my own or that there would be the slightest advantage in introducing.

Deliberately to bring into being, in any area, conditions of complete clinical sterility ranks high among my catalogue of sins and mistakes.

As I say, any bird that can manage, not only to survive this sort of treatment, but to look the picture of health at the end of it has won its spurs and can go up into the next class. Photographs exist, in Fleet Street and elsewhere, of vigorous, romping cockerels and hens, living on this piece of ground, that ought to be riddled with every nasty thing it is possible to put a name to and a few more. And anyone can judge of their condition for himself or herself.

The object of all this, and of mentioning it here, is not to demonstrate that the keeping of fowls under such primitive and seemingly unhealthy and unsuitable conditions is a good in itself. And I am very doubtful as to whether anyone else, anywhere else, could exactly duplicate what I am doing here. But it is to turn upside-down all accepted practice. The editor of a famous journal sat, one morning, in this very chair which I am occupying to type this, and posed the question: "But would it work as well with a flock of, say, 30,000?" My response—do I need to say?—was: "Nobody has got any business keeping feathered stock on that sort of scale all in one place".

Set this over against the general picture as it presents itself to those who are familiar with the contemporary poultry scene on both sides of the Atlantic and who are, moreover, able, in the words of my namesake, Matthew Arnold, to see life clearly and see it whole. The majority of

A study in devotion. Here Sister Cecilia demonstrates the importance of taking time off, now and then and here and there, from the daily round or the rush and bustle, to contemplate and converse with one's flock.

At Stanbrook Abbey the birds have any amount of room and every opportunity to live out a full and happy life amid the most beautiful and peaceful surroundings.

The Abdiction was only a few months distant when this snapshot was taken. The White Wyandottes being fed by the author's young nephew and niece were undergoing a rigorous endurance test. They failed and were reluctantly discarded.

Taking the chair. This is Rufus, the founding father (R.I.R.) whose influence upon the flock has been enormous.

farms (we again employ this term with certain reservations) both in this country and in America resemble, in this respect, nothing more closely than cities under seige. Scarcely able to sleep, either at all or without recurring nightmares in which the whole place is taken over by noxious bugs of every conceivable or inconceivable hue, shape and size, their owners or managers fall over themselves to stock up with every weapon, all the concoctions and devices that modern technology is able to put in their way. And modern technology is only too happy to oblige—or the drug-pushers and the rest of them are. Just as in the case of human ills, real or imaginary, where the policy of the medical establishment has been to continue to test out new drugs, instead of consigning the whole bag of tricks to the dustbin on this side of the Atlantic and the trash can on that, there are vast fortunes to be added to those already made.

It is heartening to note that, as conventional medicine, as practised within the U.K., continues to deteriorate in quality, and as the tendency to grab more and more in return for less and less continues, more and more sensible folk are becoming increasingly disillusioned with it. Even so, if the success of the gigantic confidence trick to which the euphemism 'Cancer Research' is applied is anything to go by, the sun looks like continuing to shine for these boys for some time yet, and a lot of hay is going to be made. One has got to take off one's hat to whoever it may have been that first thought up this, the father and mother of all wheezes—they were pretty smart. The dodge is, of course, to kid the lay public, by means of strategically-placed adverts and a number of other subtle devices, that the breakthrough may be just around the corner. This year . . . ? Next year . . . ? Some time . . . ? Meanwhile, cohorts of well-meaning if grossly misled do-gooders are roped in to go around or stand in some public place to dish out little flags, rattle collecting tins or, in one way and another, collect up the millions of pounds and dollars required to fill the gaping maw of this monster which, like the crocodile in *Alice*, sits there all the time waiting to welcome little fishes in. Many of these good if stupid people are animal lovers and in conversation with a few of them I have noted their confusion when tackled on the subject of how they can manage to reconcile this with getting mixed up in 'research' of any kind that involves the totally unnecessary and unwarranted sacrifice and suffering of large numbers of animals.

In addition to my other manifold interests and activities I have for a long time made an intensive study of human and animal pathology, and am myself a successful healer. And, so far as this dread malady itself is concerned, it can be stated without the slightest hesitation that, even after having reached an advanced stage and metastasized, it can in a very large number of cases be completely cured and the sufferer restored to abundant health. I have myself shaken hands with men and women, some of them extremely elderly, who after a succession of 'qualified' practitioners had done their worst over a number of years and found an excuse to get rid of them and leave them to die, had found their way to the right places. Cancer can be cured and has been cured in different parts of the world for generations. And, ironically, the only reason why so little is known and done about it is that, if the cure for it were to become as foolproof and universally recognised and applied as that of, let us say, toothache, a great many exceedingly lucrative pursuits would disappear overnight. There would be a rapid diminution in the cushy and time-wasting jobs going within the field of 'research'. Millions of pounds and dollars less would be raked in by those manufacturing, selling and prescribing drugs, and fat fees would cease to be come by (I was about to write 'earned', but stopped myself in time) in connection with the 'exploratory' and other kinds of butchery that go by the more innocent-sounding name of surgery.

It is, I concede, difficult for the lay mind to grasp fully the enormity of this outrage upon humanity, which is made even worse by reason of the fact that, as often as not, it sails under the flag of humanity. Even those of us who have peered, as Dostoievsky did, into the well of life, and are familiar with its seamy side, find it difficult to credit that any section of the community can be so callous and so ready to cash in on the misfortunes of others.

There need not be the least doubt that, over the years and the decades, very large numbers of perfectly innocent men and women have been deliberately sacrificed on the altar of professional arrogance and greed. They have suffered, and in many cases died, for no other or better reason than that the mystique, so painstakingly nurtured for so long, should continue to be maintained, the image remain untarnished, and the coffers at all times be kept well topped up. It never ceases to amaze me how many of one's contemporaries, who in so many other respects take up a fairly sane and sensible attitude towards things, seem to be unable to see anything much amiss with the proposition that it is quite in order for someone to die slowly and unnecessarily while under 'proper medical care', duly 'sedated' and doped, of course, in the best clinical tradition, but not for them to get better in the hands of those who have been fortunate enough to escape the handicap of five or six years at a medical school.

Every attempt is made, by those belonging within the magic circle and who are therefore able to enjoy the shelter afforded by the various trade protection societies associated with it, to damage and discredit and if need be and if possible destroy those who, given more scope, could clean up the whole fabric and change it for the better. The medical establishment has a very strong lobby at Westminster as well as a good deal of leverage with the B.B.C. and, both in this country and the U.S.A., those gentry whose machinations and whose devious ways we have just been discussing have not hesitated to use legislation, whenever the chance presented itself, for their own unworthy ends. Even if we confine ourselves to the United States alone, the story has been a shameful and shocking one, as any reader of this book who takes the trouble to trace it back can discover for himself. He can start, if he likes, with the cure of a horse in the state of Illinois during the last century and follow that up with an investigation of the curious story of one or two clinics that were closed down for the very good and sufficient reason, in the eyes of certain powerful medical interests, who were in cahoots with certain political interests as unscrupulous as themselves, that they were far too successful.

Of course what we must never, in all this, lose sight of is that in the last analysis one's health depends upon one's state of mind and basic attitude towards life. Whether it be human beings, animals or birds we are considering or dealing with, the self-righting and restorative mechanism of the body must never be upset or prevented from asserting itself and doing its work. In cases of cancer, when surgery or radium is resorted to, tissues are destroyed and this is made difficult or impossible. It has been said that a happy man never gets cancer. And I am now going to extend this and put forward the proposition that a human or hen who is happy and busy the livelong day in the right way never falls victim either to that or to anything else that need give cause for a moment's alarm. And, bear in mind, mareks disease in fowls is a cancer of the nerve endings and when they succumb to leukaemia this is a form of blood cancer.

Many years ago, walking behind a pair of Shires or Clydesdales, I used to pride myself on being able to plough a straight, clean furrow. I still do so, only it is a different kind of furrow. The late George Henderson, whom we have already had occasion to mention, for most of his life got up very early in the mornings and worked very long hours, seven days a week. He remained fulfilled and in abundant health and spirits up to, as far as my information goes, the very last. I keep on telling folk that I work a 48 hour day and a 14 day week. I sometimes despair of ever catching up with all the work I see around me waiting to be done and with no one else to do it. And there is always so much of interest and so many things to enjoy. Clock-watchers, and those who spend most of the day waiting for five o'clock, are helping to dig their own graves.

On this whole subject of the maintenance of health and vigour I should like to allow someone else to have his say. The following is taken from the book, *'Fringe Medicine'*, by Brian Englis, and published by Faber (P. 91):

'On many of the issues over which they have been at odds with the allopaths, the homeopaths have been shown to be right. Homeopaths, for example, were the first to express doubts concerning the apparently decisive allopathic victory represented by the discovery that microbes were, to all appearance, the cause of disease. In 1925 Léon Vannier, now the doyen of French homeopaths, suggested that this was an error: that in illness, 'the toxin precedes the microbe'; microbes are the agents of the disease, not its cause. This view was also expressed by James Tyler Kent. Doctors, he said,

> "will tell you that a baccillus is the cause of tuberculosis. But if the man had not been susceptible he could not have been affected by it ... Baccilli are not the cause of disease, they never come till after the disease. Allopaths are really taking the sequence for the consequence, thus leading to a false theory. You may destroy the bacteria and yet not destroy the disease. The susceptibility remains the same, and only those who are susceptible will take the disease".

—the microbes which allopaths regarded as the sources of disease are, on this theory, more akin to looters in a city where law and order has broken down. They may require prompt disciplinary measures, but there should not be any misguided notion that arresting and shooting them will save the city; that can only be done by restoring law and order—or, in the case of medicine, by reviving the life force'.

Before going on to what I want to deal with next, there is one spectre that I had better lay before it gets a chance to raise it head. Like my grandfather before me, who was a disciple of Culpepper, I have in the past used homeopathic remedies from time to time on both animals and humans. And the reference in that quotation to such practices may provoke the objection, in the minds of at least a few of my readers, that surely vaccines, whether 'live' or 'dead' (active or inactive), fall somewhere broadly within this category. I can only devote so much space here to a discussion of such matters, and will therefore content myself with stating, quite firmly and categorically, that *there can never be any justification for the use of vaccines of any kind, upon any pretext whatsoever, in connection with animals, birds or human beings*. Resort to such measures always and inescapably amounts to a tacit admission of weakness, ineptitude and failure.

The late Newman Turner of Somerset, all of whose works (also published by Faber) should be owned and never borrowed by all those having the well-being of the land, crops and livestock, along with all other created things, at heart, so that every word may be committed to memory, wrote:

'There is only one disease of animals and its name is man'.

It is going to be asserted quite boldly and confidently that every blessed ill to which domestic animal flesh is supposedly heir—not excluding foot-and-mouth in ungulates—is avoidable and is, in fact, traceable to neglect, faulty husbandry or the abuse and exploitation of the land. (In certain Swiss cantons, wherever nature's rules were strictly adhered to, the holdings concerned were never once visited by this highly unwelcome malady, which in this country has, these many years past, furnished the pretext for barbaric practices and colossal and totally unnecessary wastage, even though hemmed in by areas where it was rife). In religious terms, they can all be traced back to sins of omission or commission on the part of humans.

Some of you, reading this, will doubtless recall hearing, in Bible class, before such pursuits went out of fashion, about the chap who, on a certain famous occasion, felt prompted to inquire: "Who was it sinned, this man or his father, that this thing should have befallen him?" I think I have got the quotation right.

In Old Testament times a pastoral patriarch, of a certain cast of mind, looking out from his window or porch one morning and observing, with his practised eye, something amiss with his flocks and herds, dotting their little hills, would as like as not be moved to withdraw into his inner sanctum and commune with his soul. In what way, he would inquire of himself, have I fallen short of the trust resposed in me? How have I failed in my stewardship, that this trouble should have been visited upon me?

His present day counterpart would be far more likely to grab the telephone and put in a call to the local vet, or perhaps the nearest outpost of the N.F.U. or whatever may have succeeded the N.A.A.S., which in its turn grew out of the notorious War Agricultural Executive Committees.

No such thing exists as an incurable condition. But in these kind of situations, as in any other kind whatsoever, it is always a grave tactical error to take up, or allow oneself to be tricked or manoeuvred into a defensive position and is always the sure and certain mark of an immature or misguided person. You cannot hope to get the better of any problem or adversary running away and continuing to do so while there is still some place to run to and you have any breath left. Always stand your ground and call your adversary's bluff, whether he be a man or a creepy-crawly. The late J. W. Dunn, who wrote the famous book, *An Experiment With Time*, spoke of something he called an Infinite Regression. Like the fledgling G.P. who, quick to pick up all the dodges, dishes out a blue pill before dashing off to reduce his handicap at golf, or to dally with some not over fussy nurse or maybe some woman operating the same game as himself, only to substitute a red one when no change for the better has been noticed, and a white one when the hapless patient still complains, those clever men who, an age ago it seems now, were first called in to see what they could do about coccidiosis are still busily and expensively engaged in putting up and chasing what shows no signs of being anything but an endless succession of hares or wild geese. Like the schoolboy who has to keep on attempting to cover up the last lie with another and bigger one, *ad infinitum*, they bring out one formula, and when at last it dawns on everybody that it no longer works, if it ever did, they set to and concoct something else.

It is all one huge joke—when, that is, it is possible to forget or overlook the tragic side of it and how stupid it is. There exists, in fact, a close parallel in the efforts of those other bug-hunters who operate on both sides of the Atlantic and pursue aphids and all those other tribes of creeping, climbing, flying, boring and champing creatures that take a close interest in plants, foliage, fruit and such. When the aphids or whatever first built up an immunity from the preparation being dusted or sprayed or otherwise applied to their chosen habitats, something different or more potent was tried, until now the process bears every sign of going on for ever and there are little chaps around today, acknowledging both American and British citizenship, who lap up with evident relish stuff that would have killed their not all that remote ancestors stone dead.

We are, then, confronted with this most curious dichotomy; with a situation in which, on the one hand, those who have graduated with honours in public relations and both the hard and the soft sell really get down to the job of persuading the ones who are still running in blinkers that the pullets, broiler chicks and the rest churned out by the massive concerns are the cat's whiskers, and on the other a constant succession of curses, wails, and cries of despair issue from my telephone receiver.

Ever since Henry Wallace, the American politician and maize breeder, first invented the hybrid millions of dollars and pounds have been spent in an endeavour to persuade those who have not already had their fingers burnt that it is God's gift to man. Yet never a day goes by but some man or woman who is poorer, sadder and, in short, sick to death of having to perform last rites and organise burial parties calls me up from some part of the world to ask if I am able to provide him or her with something that looks and behaves like a hen and can stand being outside in the fresh air and rain without going into shock, in addition to knowing what grain and grass are for. Over and over again letters reaching this office contain the phrase '. . . they die like flies when put outside'.

One Sunday lunch time, during the last war, in an American officers' mess, they placed before me a heap of knobbly things on a plate. When I looked at it and then looked at the Servicewoman who was acting as waitress, she vouchsafed the information: "It's braised chicken". This, though contributing somewhat towards my education, did nothing to allay my suspicions. It might have been anything, but if it was meant to be chicken of a sort, where and in what set of conditions had it begun life? More important still, where and in what circumstances had it met its end?

I hold that any creature qualifying for description as a chicken, whether alive or dead, and at any rate until the third day and the soufflé stage has been reached, ought to be recognisable as such. In its complete lack of identity the modern hybrid corresponds to those plastics which, neither fish, flesh nor good red herring, have so largely supplanted honest woods and metals.

For a pullet to be deliberately produced which has no body size, so that, when dead, even if not riddled by disease there is nothing on it, and that, moreover, can only be kept alive by being artificially propped up with antibiotics and one thing and another, is just crazy.

On this farm, if a bird should cut its foot on a piece of broken glass or stand on a felt nail—a very rare occurrence indeed—we bathe the wound and put Eusol on it. Beyond that, no birds is ever treated or isolated. Louse powder or any such preparations are unknown, and we never see hide nor hair of such creatures. I have never fumigated an incubator in my life. Disinfectants and such things as red mite repellents are banned from the place. I do not even use wood preservatives; and the only red mite I have ever seen on the premises appeared in a small wooden appliance that had been purchased second hand, having previously been creosoted.

Which, I think, may be said to tidy up that part of the discussion reasonably well, or as well as we have space for within the compass of a book and a chapter such as this. All that we have been saying naturally ties up with the important question of longevity. One day, some years back, at a place not far from Blandford in this county, I was invited to look inside a stable belonging to a military man. Having gone over the two flat racers standing there, and noted their conformation, I announced: "These are middle distance horses". Their owner nodded his assent. It is not the intention of the author of these pages to fill them up with masses of figures and statistics or even with information that is as readily available elsewhere. But anyone having access to a reasonably sharp pencil and scrap or two of clean paper should not have much difficulty in working out, given the present day cost of the various items involved, the approximate difference between taking on board a batch of pullets each of which is geared to no more than a few months' lay—if they can by one means or another be kept alive and in production that' long— and one every member of which has inherited the capacity, under favourable conditions, to turn out a profitable number of eggs over a great many years.

No owner or trainer in his senses would dream of entering even a middle distance horse in a race like the Ascot Gold Cup. Much less a sprinter. The modern hybrid falls into the last classification. And it does not require a mathematical genius to see that the difference between having to bury and replace a whole lot of carcases every few months and having something worth putting in the pot at the end of eight or ten years must be enormous.

For years there stood in a corner of our kitchen here a very large commercial 'fridge of a famous and reliable make. It was solid and workmanlike and looked as if it would last for ever. Unfortunately, a few months ago, it happened an accident and we found it impossible to replace the damaged parts. So, very reluctantly, we had to lay out cash on the best substitute available in these times. Over the past decade or two, both in America and this country, there has crept into the nomenclature associated with such things as motor cars and refrigerators a new expression: built-in obsolescence. Now it is a complete waste of time, labour, cash and everything else to manufacture and place on the market any article soever that will not last, usefully, for as long as anybody knows how to make it last. To argue that deliberately to turn out products that will soon need to be dumped and replaced is a good thing in that it creates more employment is sheer bunk and eye wash.

Whether or not those wide boys, some of them known personally to me, who are based in California, Canada, here and in Europe and who turn out the sort of rubbish we have been discussing like tin tacks from a machine deliberately tailor them for a short life or whether they couldn't do any different if they tried, we need not discuss here. That is certainly the effect. And what we can safely bet on is that, gazing upon the works of their soft, white hands, they rejoice and are glad.

CHAPTER THREE

THE KINDERGARTEN

I must confess to having experienced some difficulty in deciding how to begin this chapter. I do not mean knowing what to put—there is never any trouble about that. Indeed there is so much that one could say, if let out on a long rein, that the problem at all times is to decide how to compress into a single paragraph what could easily run away with itself and finish up as a couple of pages or more.

No, what I am referring to is the particular stage in the avian saga with which this section, bearing the label it does, ought to deal first. I should have liked to have begun with the baby chick: those sweet, fluffy little things which tend to draw from young and old alike—those who have not yet lost their innocence and become case-hardened or are not going to—such exclamations as "The little darlings!" or "Oh, aren't they lovely!" But if I had responded to this impulse there would have been no logical or natural place throughout the rest of the book at which to say something about incubation.

When the subject of the commissioning of this work first came to be discussed, the publishers asked me for a synopsis. I replied that this was a task that lay beyond my capacities. Once the job was completed and I had a chance to take a good look at it and see what it was all about, I might perhaps be able to rise to it. In the meanwhile . . .

In the event they settled for a compromise and I was able to manage some chapter headings, as I do have a rough idea in my head as to the final shape of the work and the general direction it is to take. And as I say there is nowhere else for the pre-chick stage to go.

Let us begin, then, with a word or two about a process that will ring a bell in just about everybody's mind, expert and layman alike— the act of mating. No egg is, of course, of any use except for eating until male and female have been united in this act. What takes place, very briefly, is that the reproductive cells of the male are introduced into what is known as the cloaca of the female. From here they make their way to the top end of the oviduct. When an ovum or yolk enters the oviduct one of the spermatozoa present joins up with the female reproductive cell in the blastoderm of the yolk. What we end up with is called a zygote, which means yoked together. It is, therefore, at the time of mating that the sex of the resulting chick is determined.

From this it will be seen that the hen does not have to be mated afresh for each egg she lays, although a vigorous cock will often mate with the same hen or pullet over and over again even when there are other candidates around. In fact, for most practical purposes, the female can be allowed if need be to remain a grass widow for about a week after the mating. Not many fertile eggs will be gathered after that, although there are exceptions to the rule and progeny has been known to show up at varying lengths of time afterwards, occasionally up to three months or so, which clearly resulted from a previous mating. This is known as telegony. And, of course, the same thing applies in cases where one male is substituted in, so to speak, mid stream, for another. In the ordinary course of events chicks hatched from any eggs laid after about one week are to be credited to the new one.

Hens' eggs, as distinct from those of certain other species, take 21 days to hatch. Sometimes they never hatch at all, and the reasons for this are numerous. Quite apart from infertility, an egg can fail to deliver the goods because of deficiencies in the diet of the hen that laid it, or the conditions under which she has to carry out her labours. It can remain unhatched as the result of conditions in the incubator house or incubator itself, or of factors having to do with temperature, or moisture or lack of it.

But—and here is one in the eye for all the clever dicks—if a hen is unable to bring off a stale or difficult egg, it is a waste of time and an instance of grossly misplaced confidence to expect a machine, however sophisticated or finely adjusted, to do so. I read somewhere that the total functions of that marvellous piece of mechanism, the human eye, could not be duplicated, if at all, by a computer occupying less space than a full-size piano. The biggest and most advanced incubator money can buy hasn't got a hope of coming within hailing distance of the humble and much-abused hen. Its only advantage—if it is an advantage—lies in its ability to take in and regurgitate a much greater number of eggs.

I cannot remember the last occasion on which I broke a broody by placing her in an outside cage with slatted floor, so that it was more uncomfortable for her to sit down than to remain standing. We do have plenty of broodies out in the fields, but they are allowed to sit for as long as they have a mind and in fact most of them are roped in to bring off chicks and rear them as nature intended, as well as keeping down the level of unemployment. Not many days before this was written one of our Dorset Reds proudly presented us with twelve out of thirteen. Thirteen being the ideal number for one bird to take care of, unless she happens to be an unusually ample matron, in which case you can perhaps entrust her with two more. A clutch numbering twelve or fourteen won't do. This, I assure you, is far from being an old wives' tale, for 'clockers', as we used to call them up north, have a strong dislike of even numbers and will often deliberately break the extra one, not 'just to even things up', but to do the opposite, if this is tried. Nature, all the way through, it must be remembered, exhibits strong preferences when it comes to numbers.

A hen will do even better, most times, if left to make her own arrangements and better still if, having stolen her nest and sat it out in a hedge bottom or some such place, unmolested by foxes or other predators, she proudly, her task completed up to that point at least, struts in the direction of the farmhouse, bringing along her troupe for inspection.

It may be thought that a pedigree breeder such as myself, with enthusiasts in many countries sending in orders and inquiries the whole time, ought not to feel it worth his while to involve himself in such capers, despite the enormous pleasure to be obtained therefrom. But the broodies are not looked after by me personally, and it seems to fit in well enough with all the other jobs. Indeed it is not unknown for one person to look after a very large number of sitting hens. In our case the chicks which result are primarily intended to swell the ranks of our own breeders when big enough, and not to go out to customers. An old and valued friend of the author's, Jim Worthington, expressed the belief that it was essential, if one's stock was to be maintained at a high level of vigour and performance, to keep on going back, at frequent intervals, to mother nature. And it is indeed difficult to sidestep the conclusion that, as with Napoleon encamped before Moscow, it is an easy matter for us to allow our lines of communication to get too long or tenuous, so that they are in danger of becoming breached here and there.

We can, then, learn a great deal from the hen if only we are sufficiently humble and exercise enough patience and observation; for the overweening never acquire wisdom or understanding: from the way she, cunning old soul that she is, goes about her business, turning the eggs, adjusting the temperature, humidity and so on.

The bulk of the eggs go into a row of Ironclads that I installed, when first purchased, on a long bench. Each of these is intended to take 200 hen eggs, although this depends on their average size and most incubators will take a few more than is stated on the literature. The bench, constructed out of pinewood that dates from the time when timber used to be seasoned before being sold or used, has its upper surface exactly 2ft. from the concrete floor. There are those, incidentally, who believe that an earthen floor is preferable for this purpose, as it cuts down the number of dead-in-shell; and I have at one time and another used both. Otherwise the Ironclad is not all that particular and all it asks is a roomy, airy shed.

To continue. The depth of the boarded top of the bench—that is, from back to front—is 3ft. 2in. The minimum space that can be allowed for each machine, complete with oil tank, to occupy is 4ft. And each incubator measures approximately two feet to the highest point.

I prefer a paraffin-heated machine to an electric one; although the makers, Messrs. C. A. Sydenham-Hannaford, Ltd., of Hamworthy Junction, Poole, Dorset, turn out and export both kinds. The founder was himself a great lover of fowls, and kept one or two of the beautiful old-fashioned breeds. So many great or successful businesses, standing the test of time, have grown out of a handful of fowls, beautifully tended, in a garden or backyard and kept for eggs or exhibition. Quite frankly, within the range of capacities embraced, I do not see the advantage of going in for any other make. In saying this I had better stress that it is nothing either in or out of my pocket what you invest in. I receive no payment or concessions of any kind for recommending the products either of this firm or any other; nor would I accept any.

But let us not get ahead of ourselves. Let us, instead, go back, lest such an important matter should be overlooked, to a consideration of the nature and quality of the dough which should be fed into this particular kind of oven. Even if your cockerel and his harem are all fertile and all behaving, throughout the livelong day, as nature intended and you, their proud owner and attendant, fondly hoped the battle is not yet won and this is only part of the story. A rural stonemason whom in my youth I knew was fond of declaring, to anybody who had not heard it before and had the patience to listen: "What I say is—one man's as good as another, and in fact a sight better". One egg isn't as good as another and not all eggs are fit to go into an incubator or under a hen. If they have been collected from the nest at too long intervals, particularly during extremes of weather, or improperly stored or otherwise ill-treated in the meantime, it will not be worth while risking setting them. In the course of my own experimental work I have successfully hatched eggs, in old-fashioned wooden incubators, that were up to six weeks old and had never been turned. But in ordinary practice they ought not to have been laid for more than ten days. Whilst in store they should be turned morning and night. Obviously the precise manner in which this is carried out will depend upon the scale of operations and one's own special or general circumstances. But for those who have only a small number to deal with although, large farmer or backyarder, all eggs intended for hatching should at all times be stored on their sides and never end-downwards, the simplest and least time-consuming method of accomplishing this is to install one or more rigid trays constructed out of close-mesh wire. These are propped up on a convenient bench so as to provide a gentle slope; and the idea is to begin filling them up towards the top, the eggs being checked in their attempt to roll all the way down to the bottom in one go by a suitable piece of wood. This is gently moved downwards a little each time, in fact just enough to enable every individual egg to move to a fresh position. What we do on our scale is to pack them into small, shallow orange crates with open sides in between layers of the short stuff that falls out of the wheat straw when we thresh it for thatch. We count exactly 100 into each of these and secure the lid tightly. After that all we have to do is to turn the thing bodily at either end of the day.

All eggs not suitable for such treatment must unhesitatingly be consigned to the kitchen or market. Those having thin shells, or blemishes or wrinkles all over them, are discarded. The first of these problems can generally be overcome by the provision of grit. Depending on where they are situated and what lies underfoot fowls can sometimes pick up much of their own requirements in the way of soluble grit for egg-making or insoluble, in the form of granite, flint or whatever for grinding purposes. But most have to be provided with at least some of one sort or the other. Naturally the call for the soluble is much the greater; and I prefer hard limestone to soft, and find the kind and size which one sees spread on road surfaces ideal, with a little Scandinavian oyster shell to help harden the shells.

On the subject of the various other defects we have listed, I am reminded of the remark of the chap who said of a particular horse which, though excellently bred and sound in wind, limb and eyesight, nevertheless failed to come up to his owner's expectations, that he would be a good enough horse if only he had a good enough driver. For me to unpack the eggs and come upon one that for one reason or another is not good enough to set is all but unknown. The ideal hatching egg is a well-shaped one having plenty of girth at the middle, to allow for the proper development of the various organs, and weighing somewhere between 2oz. and 2½oz. These two conditions having been fulfilled, the shinier it is the better. Some of ours, admittedly, have less polish on them than others. But that having been said it is quite impossible to over-state the importance of the role of the person whose finger is on the pulse of the whole show. Those, concerned with crops and livestock, having a negative outlook and approach, tend to produce results which are negative when weighed in any valid balance, whatever their bank statement may show. Mis-shapen eggs are produced by fowls belonging to or managed by mis-shapen people, or by those which have been bred and reared by those having mis-shapen minds and twisted characters. 'By their fruits shall ye know them'.

So we have got the eggs ready to be placed on the tray which is made so as to slide on a runner fixed to each cheek of the incubator in the same way that a baker slides a shelf containing raw loaves into his oven. The Ironclad is well insulated, being made of steel, and so it is not difficult to maintain a constant temperature. It is run at 104° for most of the time, and I am so used to mine that I am quite sure I could bring off a successful hatch without the use of a thermometer at any stage. Perhaps I will do it one day, for a bet. You, of course, dear beginner, will have to learn to creep before you attempt to walk, just in case you happen to fall down. You will have had all the 'innards' carefully put in their place and the machine running at the correct temperature for some hours before trusting those precious eggs to it; but of course the cold eggs will bring it down, and it will take quite a while to recover.

Most of what else you need to know about incubation can be obtained from the makers or other books. It is usual to test all the eggs for clears on the 7th or 10th day, but I wait until nearer the end of the hatch before removing any. Curiously, as the excitement mounts and the moment of truth looms nearer, it matters less what happens to the temperature—so long, that is, as it happens in a downward direction. In all these years I have been at it some queer things have been witnessed. I have, for instance, watched perfectly normal chicks in great numbers emerge from an incubator that, according to every law anybody could think of, should have yielded nothing but charred remains. All the same, and despite the saying to the effect that fools walk where angels fear to tread, you would be well advised to play safe and to proceed on the assumption that incipient chicks are not improved by being roasted. No, what few if any can understand is how it comes about that, once it gets to about the 19th or 20th day, eggs can often be exposed without harm to the most bitter weather conditions. I have myself often seen the most wonderful hatches follow upon a period during which all the eggs were to all appearances stone cold and every last chick ought to have been stone dead. A woman living on the borders of Devon rang me up one morning to say that, for the first and last time in their joint lives, she had gone out for the evening and left her husband to see to things. When he switched off the light in the aviary he inadvertently followed suit with the electrically-heated Ironclad. We were in the depths of winter and she was very much inclined to throw the lot out and start afresh. I asked her how far the hatch and progressed, and when told that it was near its end, urged her to do nothing of the sort. I am sure she thought I was mad. Anyhow she carried on and a few days later I received a telephone call from her in which she said it was the best and biggest hatch she could ever remember. She and the forgetful one were now reconciled and in future she would make this a general practice.

When you receive the instruction sheets on the Ironclad it will be seen that, with this machine, as soon as any eggs at all are found to be pipped, the water trays are removed and the thing closed. It is not opened again until 48 hours later, when the chicks are all taken out and it is allowed to go cold. In any case, before closing it up, I remove the tray and rest it on the trolley I made for such purposes. All the eggs are tested and any infertiles removed and placed in a bucket. We shall see in a minute what happens to these. But the testing is done in a darkened corner, each egg in turn being looked through while a strong light is held behind it. Some operators use a torch for this purpose, and others rig up a box with an egg-shaped hole cut in it and containing an electric lamp. A clear egg shows up like a newly-gathered one. One of the objects of testing them at an early stage is to enable, where possible, those remaining in, say, a dozen machines to be accommodated in about ten, thus freeing the other two for a further batch.

Of course geneticists and pedigree breeders such as myself need to know a lot more about what is happening inside the eggs at different stages than you as a beginner need to concern yourself with.

It is at this stage that those eggs remaining which belong in special categories and have to be kept track of are fished out from the rest and placed in pedigree bags or cages. To make it possible to identify them they have already been marked—in fact at the time of collection to avoid any possibility of a mistake or mix-up—with a black wax crayon that will not wash off. Not, mind you, that you should ever wash hatching eggs, however dirty. But they might get rained on, or if anything less durable were to be used it could rub off. Every breeder has his own set of symbols. The system I have for years followed here is quite simple and, once got into, proceeds like clockwork and saves a great deal of work and confusion. I have a separate letter of the alphabet for each successive batch or hatch. This follows the whole lot through from birth to death, or from the egg to being sold to a customer or customers at some stage or other. In the case of a chick to be permanently identified and kept tabs on for any reason it follows it wherever it goes and is, indeed, the very thing which I and a number of others are actively engaged in trying to prevent happening to human beings in this country as it has to so many millions in other parts of the world.

In fact each egg and the chick resulting from it has to carry, not only the batch index, but also the index belonging to that particular individual. As an illustration, all the eggs laid by one brown hen, mated to a brown cockerel, sport this mark: ☐ This is placed at the blunt end and also the sharp end. In the case of a mating between two different breeds or strains, the cockerel's symbol will appear at the blunt end and the hen's at the other. The batch index is to be found on the fattest part of the egg in three different places. This is so that it can be spotted at a glance in the incubator or wherever without the necessity of turning it over.

All this information goes down on a special card that is pinned to the bench in front of each machine in the centre. This card also tells us on what day in what year and at what time the eggs were set, how many of each kind there are and, in short, everything we need to know about them. I have never used the old-fashioned wing-bands for years; not, in fact, since before the war. The small wing-tabs in favour now are clipped on to the conveniently loose piece of skin that is to be found just above the chick's 'elbow'. This causes it only minor and temporary inconvenience and it is very rare indeed for one to come adrift. Each bears a code number for which no duplicate exists anywhere in the world. The practice followed on this farm, however, is at variance with that pursued elsewhere and recommended by the makers. The tabs are not affixed until the bird is at least ten days old. We feel that it is too heavy to attach to a newly-hatched chick, and it might get caught in some projection. So, whenever any of the babies have to be kept track of—which is after every hatch—my wife ties different-coloured wool to the base of their wings. My fingers were not meant for such delicate jobs as tying wool or threading needles. One colour—let us say Air Force blue—goes on the left wing of all chicks belonging to the same batch; and a different colour for every separate chick or category is placed on the right wing. When the time arrives for the fixing of the tabs it is a simple matter to sort out which is which and book them down as they are done.

The final act in this sequence consists of hanging a weatherproof letter on the brooder or piece of rearing equipment containing the batch, or what is left of the batch to which that letter refers. In this way it is child's play to check up on any lot of birds at any time and to know, not only exactly what age they have got to, but where their dossier is to be found.

All this having, one hopes, been duly assimilated, we need not contain our impatience a moment longer, but can go back to those fluffy baby chicks that, when we left them, were in process of being taken out of the incubator. To a real stockman, as distinct from those hard-faced characters who peer out at us from the pages of the glossy magazines allegedly devoted to poultry husbandry, this ever-recurring miracle of incarnation is a never-failing source of wonder. Once the little chap has dried out, particularly if it is a big one—and some of ours are like stags and, given half a chance, are off across country at full gallop before anyone has time to stop them—even the most experienced of us are at a loss to make out how it ever came to be contained within that egg. But, by a miracle of nature, contained it was—an amazing bundle of energy and potential.

The making of a new breed. Here is shown a small part of the original hatch of Dorset Reds and Whites at a few days old.

These up and coming Dorsets, committed to the tender mercies of Sister Cecilia, give every indication of being contented with their lot. They also help to bear out the advantages claimed for the Golden type brooding unit when combined with electric heating. (Observe the two pairs of cables.)

Something to look pleased about: the original Dorset White hen—dubbed by David Wilson 'Grandma of them all'—is held up for inspection by her breeder.
Photo by courtesy of *British Farmer & Stockbreeder*

We have already drawn attention to the difficulties encountered when trying to convey to the student or other interested reader or listener what differentiates a good bird from one that is not so good, or no good at all. And what applies to older stock applies equally to the youngest of all. It can be stated, for a beginning, that the sort of chick which ought to be looked for is vigorous, active and alert and has a bright, bold eye and something to say for himself or herself. If the 'hair' on its head appears to have been carefully brushed and flattened down, you may be on to a winner. Listless ones and those that seem to look upon the world as an unwelcome place are neither use nor ornament. It should be well-balanced, with a pair of sturdy legs, and stand four-square.

But we must not keep them hanging about, must we? At this stage such little creatures as these, robust or not, are easily chilled. So into the railway boxes with them if they are going away and into the brooder with them if they are not.

When being despatched by rail day olds are packed in special boxes lined with soft, sweet hay to act as insulation and to help conserve their body heat. Those we use can be adjusted to suit just about any climatic condition, from arctic to tropical. And, whilst on the subject, once in the box they must never be placed near heat. The smallest number any box is normally made to hold is 25. So if you want to send away or receive a smaller number than this, somebody needs to make a smaller nest. Even then it is risky to send baby chicks any distance in cold weather if the box contains no more than around a dozen.

Let us now turn to the important subject of brooding, to which I have given so much attention over the years, having at one time been, in addition to looking after my pigs, birds and other stock, a designer, manufacturer and distributor of a wide range of poultry houses and appliances.

The design of brooding unit which I have in the end come up with, and found after exhaustive trials highly satisfactory, I call the 'Bradstock'. There is an indoor and also an outdoor version, the latter being moved daily at the end of an electric cable. This last enables chicks to go straight from the incubator on to grass, the whole thing being surrounded by a seal to keep out draughts. Floor draughts, never forget, are fatal to your pets at this stage in their career. It will be noted that the heating units specified incorporate ordinary light bulbs. On no account should infra-red appliances of any kind be used for the brooding of chicks. Ducklings, yes.

It had better be made clear, before passing on, that a brooder of the length and width of that shown in the constructional drawings is not going to accommodate many chicks even up to a fortnight old. Both can be extended to suit individual requirements so long as the proportion remains the same and no alteration is made in the basic design or that might affect the function.

The next thing upon which we are going to place an embargo is the feeding of what are sold as chick crumbs. In fact, if you have taken care to get your chicks, or their parents if you have bred them, from the right place all ground up, pelleted or otherwise concocted or dished up mysteries in bags should be avoided like the plague, quite independently of the fact that they represent a sheer waste of money. If you cannot prevail upon your charges to thrive and feather and grow up as they ought, and continue to fill the egg basket once they have, without recourse to such temptations, then they were a bad bargain in the first place. If the chicks are being brooded indoors, because conditions are such or your situation is such that to do otherwise would be difficult this should not be continued beyond the third week. By then they should be outside. But while they are inside they should be supplied with a little coarse sand from which they will in time graduate to larger and larger sizes of insoluble grit. Otherwise their diet will consist of three parts by weight of the best and plumpest oats you can come by, Scottish for preference if you happen to be resident in the U.K.; Australian if you are nearer there; mixed with one part of barley and rolled. Not all the little local millers have as yet been swallowed up by the big ones, and most poultry keepers who are not in a big enough way of business to install their own mill are able to have this prepared for them. But, whatever you do, take care never to be fobbed off with what the miller or merchant 'recommends', or 'our own special corn mixture'. He is not likely to be as unscrupulous as some of the big compounders who have a habit, whenever anybody offers them a shipload or two of spoilt American herring or some such thing for next to nothing, of suddenly deciding that this is the best thing for their clients, or their clients' stock, and switching the formula to take it in without letting on. All the same, and quite aside from the fact that it is the last thing you want, it usually works out more to his advantage than yours and always puts temptation in his way whenever he has got a big heap of some particular thing on his hands and his mind that he wants to get rid of without loss.

The oat, as a valuable plant for all those making their living, or part of it, from the keeping of fowls, has never, up to now, received a fraction of the recognition that is its due. And wheat has been very much over-rated.

Once all the infertile eggs—now you see the wisdom of leaving the testing until a late hour—have been cooked, mashed up small with a steel potato masher and fed to the babies, having been well mixed into about half the quantity of the oats and barley, and this first feed has been duly scoffed, the corn is fed dry. To it should be added cod liver oil at the rate printed on the can. This comes in one-gallon containers unless you are in a bigger way of business than most; and it is usual to rub it well in by hand and then give the whole mixture a vigorous stir. The scheme here is to place a quantity of whatever is available on hatching day just outside the heated end of the appliance and stand one or if need be a couple of small fountains near it to get the inmates started eating and drinking. Once they have got into the way of things their stamping ground can be extended. Some people, I know, go to endless trouble over them and only stop short at fitting each out with a bib and tucker. They also box them into a confined space to begin with, letting this out gradually as they find their feet. However I proceed on the principle that chicks which need to be fussed and coddled, and haven't enough sense to know what good grub is when they see it, and to wade straight into it from the incubator, are not worth keeping anyway. So what we do, then, is to set a drinker having a deeper lip and a large enough capacity to fulfil the needs of that number of chicks as they continue to grow on a couple of bricks, to keep it clean and up out of the litter, at the far end of the cold section. This is to give the little chaps something to do and represents their first lesson in working for a living.

As for the crushed corn, a place is swept clean on the floor near the front and a sufficient quantity of this is spread out there. Beginners keep asking me why they cannot be tidy and confine it within one or more hygienic metal feeders. And I have to explain that the idea is for the chicks to pick out the kernels and leave the fibrous husks. These husks would quickly bung up any hopper and a famine would ensue.

At the same time as this we take the trouble to weigh out a quantity of white fish meal so as to add and mix well into it $1\frac{1}{2}$ per cent limestone flour and $1\frac{1}{2}$ per cent steamed or sterilised bone flour. This goes into one of those galvanised hoppers 18in. long and having a top which slides on and locks. This cover also has a series of holes along each side to enable the occupants to put their heads through but prevent them from entering bodily. Raw greenstuff should be provided from about the third day onwards. As soon as it has been gobbled up or has lost its freshness it should be replaced. Large lettuces are best for this purpose. Dandelions come next, and salad burnet is a very useful plant to know about for stock of any age. Care should be taken in feeding to chicks grasses having long, tough stems. When they are outside it matters less because the grasses are anchored by the roots. Or, of course, if you or they prefer it or it is found more convenient, there is no reason why you should not drop in a nice, close-cropped turf or two.

Let us now move outside and deal with the babies down the paddock or occupying the brooder in the orchard or on the lawn. Those who for one reason or another cannot or prefer not to install one or more outdoor brooders of my design need not lose any sleep over it. Outdoor brooder units of the type originated by W. M. Golden have been in use in many parts of the country for a great many years and thousands of sturdy birds have emerged from them to go on to the next stage, although I would be happier if I knew they were all heated electrically. My good friends at Stanbrook Abbey, Callow End, Worcester, have for years employed similar arrangements for their Dorsets with great success. A beautiful and extensive orchard has been set aside for the purpose and the presiding genius here is Sister Cecilia, who is completely dedicated and keeps and supplies me with meticulous records.

It has been pointed out that, if the chicks pull the heads off or pick the seeds out of a few long, anchored grasses they will not come to any harm. But it must not be thought that the young of the fowl, any more than that of the turkey, can be turned out amongst long grass and nettles and things of that sort with any hope of keeping many of them alive and well. The grass should at all times be kept as short and sweet as possible. I do not propose, here, to launch into a lecture on grassland management. André Voisin and others have made valuable contributions to this; but land on which fowls of any kind or age are run ought never to be allowed to become acid. And if great drifts of wild white clover are to be seen here and there, so much the better. It is sometimes less expensive, time-wasting and heartbreaking to learn from the blunders, mistakes and misfortunes of others than our own. If it is the same mistake, there should be no need for more than one person to make it, so long as the grapevine is in good working order. The late Richard Rodwell, it used to be rumoured, lost around £20,000 in one fell swoop through allowing the grass in his breeding pens to get long and coarse. It provided an ideal place for intestinal worms and parasites to flourish, with disastrous results to a large proportion of his valuable birds. He had occasion to write to the author shortly afterwards and his rueful if characteristically philosophical post mortem judgment was: 'Perhaps it was as expensive as that. But it has been a great fight and I have enjoyed the scrap'.

Incidentally, Rodwell, a man of the humblest origins, came to be dubbed 'The world's greatest breeder of utility poultry'. Odd people here and there have been overheard saying that his mantle has now descended upon the author of these words. However the becoming modesty for which he is so well known precludes any comment from him.

Now the extent to which you feed and treat these outdoor birds in the same way as the indoor ones must be left to your own judgment and discretion and will depend largely upon the time of year, state of the herbage and so on. If the grass is of a good mixture and palatable and there is an abundance of organic life about it should be possible to dispense both with the c.l.o. and fish meal. Cod liver oil was only introduced, in the early days of the trying out of different artificial brooding systems, because of its anti-rachitic properties. Many chicks, brooded intensively, were found to develop rickets and troubles of that sort.

Right. If you have benefited as you ought from the advice proffered, and brought the little darlings along satisfactorily, by the time they have reached four or five weeks they will require a lot more space—for the tendency of chicks of the right antecedents and upbringing to grow out of things in a surprisingly short space of time must never be lost sight of—and a slight change of scene. In my own case all chicks, whether being run on for customers or reared for home use are moved, at three weeks, to a range of structures where they can run in and out of a sort of cubby hole with slats covering the floor. A layer of soft, clean hay is spread over these, and one new 100 W. lamp is screwed to the roof about the centre. These are, of course, connected to a cable which runs out from the main buildings. Gradually the hay disappears and is not replaced. The lamp also is allowed to go out after a period whose length is determined by the weather prevailing at the time.

In view of the fact that so many excellent designs of the various carry-on appliances needed have, over the years, been brought out and published, and the further fact that we cannot afford to let this book simply grow and grow until it is as long and bulky as *'War And Peace'*,

I do not think a lengthy discussion or detailed description of my own methods and creations would be justified. These are adapted to my own peculiar needs and circumstances and it ought to be sufficient to mention that the growers remain on slats for some time and eventually graduate to perches.

Out in the big world and away from this farm the favourite ploy is to house them in movable folds, of which there are many variations, either until they are ready to go into the laying houses or until such time as they may be considered reasonably safe from cats, hawks, dogs and other predators. This is usually fixed at round about ten weeks. After that it is common to see them occupying rows of small slatted-floor arks.

What it is essential to deal with here is the matter of what food to give them and when. Once the birds have left the brooder my practice is to tip out one or more scoopfuls of the crushed corn, according to their number, on to a removable wooden tray. This has sides all round three inches in height to eliminate waste. It is left before them all the time, and before a fresh lot is put down, as in the brooder, the husks and leavings are swept up and put in for an older lot to pick through. It is a bit like the hand-me-down ritual that used to be gone through, often with surprising success, in households containing large families but not much money. When the children of one sex came, as my mother used to say, 'in steps and stairs', as the biggest grew out of, say, his jacket, instead of hanging on to it until his elbows protruded and it was of no use for anyone else, the next one down would inherit it, and so on. This habit and other, similar ones, I learn, are now spreading to other classes, what with inflation, the various brands of political and economic lunacy there are about, and the persistent attempts of the state to get its greasy finger into every pie there is going and obtain a stranglehold on even the smallest and most fiercely independent business. Only, of course, in the case of me and my chicks it is more of a 'hand-me-up' process.

After being 'weaned', and until the growing stock reach eight or ten weeks of age, they should be fed morning and night as well as at midday and allowed what they can put away inside half an hour. Up till then they may as well be given access, *ad libitum,* to both whole and crushed grain of the same mixture, fish meal being optional and depending on the conditions already mentioned. From around ten weeks, irrespective of the state of the grass or anything else, and at least until they are within a couple of weeks or so of beginning to lay, they should receive nothing but the whole corn. To give an idea of the amount to put out, adults in lay are allowed $1\frac{1}{2}$ oz. per bird, scattered about in the grass first thing each morning. Between then and about an hour before dusk they must find their own or do without. When that time arrives they are given access to a series of rat-proof hoppers, which are closed at shutting-up time or after they have eaten their fill. In this way there is no need to have to calculate, each time, as so many do, how much to dole out, and no bird need climb the stairs hungry as the result of human error. During the stage we have just been dealing with the growers can be treated in much the same way, adjusting the amount to be broadcast according to their age.

No pullet or hen that is worth house room should have to be fed protein in any form unless the grass is poor or frozen. I have had 100 per cent production from August-hatched pullets during the early months of the year on corn and nothing else. Our breeders here are sometimes given a wet mash once a week as a special treat. And for this purpose a little of the crushed corn may be used. Otherwise that is all they get.

At the tail end of this chapter I am going to lay it down as an immutable law that, *if a hen is to live to a great age, remaining healthy and filling the egg basket year after year, its diet must consist wholly or almost so of whole hard grain.*

CHAPTER FOUR
SOME BREEDING SYSTEMS, AND IN PARTICULAR MY OWN, EXPLAINED

Charles Baudelaire begins his famous story, *'Tristan and Iseult'*, with the announcement: 'My lords, if you would hear a high tale of love and of death . . . '.

I bring you, on the other hand, a tale of love and of life—of life burgeoning, and swelling, and filling the cornucopia to overflowing.

It so happens that I was nurtured in the middle of one of the world's great cradles of stock-breeding; and in fact the farm on which the original Shorthorn first saw the light of day was almost within walking distance. I am descended from a long line of aristocrats who over many generations bred and raised fine horses, cattle and sheep. And by the time the mid-'thirties arrived, in addition to my experience of and skill with these three kinds of stock, I had already gained a reputation for 'having a way with hens'.

My commercial layers, distributed over a number of herby pastures and meadows, simply poured out eggs. And on the breeding side a number of exciting and profitable discoveries were made, and lessons learnt, at an early juncture in the proceedings. My first choice of breeds was the Rhode Island Red, with White Leghorns as a second string. The beautiful White Wyandotte, to the fore right up to a year or two previous, had degenerated and gone out of favour.

In one, bold experiment I purchased a mated pen of a breeder who operated from a high-lying farm in the West Riding. He spent a good deal over advertising and much emphasis was placed upon the egg records set up by those hens from which the day olds and other stock on offer had allegedly or in fact been bred.

One day, I remember, an intrepid and enterprising young female reporter from a Leeds newspaper dropped in without prior warning. No doubt this placed the proprietor in a terrible dilemma. For if he had responded to what is likely to have been his first impulse, and thrown her out neck and crop, this without any further comment would not have made the best of reading from his point of view, or done anything to allay whatever suspicions may already have been in currency. On the other hand . . . In the event she was able to effect an entry, and she told of how her eagle eye, trained as it was to miss nothing that might come in as grist for the mill, had spotted dead and dying birds here and there in odd corners, and others which could only just stand up.

To breed from egg records and nothing else is to begin at the wrong end. It was about this time that what was known as the 'copper ring' hen became popular and then notorious. This was a pullet that had delivered 200 eggs within a specified number of weeks at certain of the big laying trials. When placed by its new, fond owner in the breeding pen it was found, in the great majority of cases, quite incapable of passing on to its offspring any qualities worth writing home about. Not everyone who entered birds in such tests was as honest as he might have been and the buying public had the right to expect. And I have known cases where the winners, like some of those sheep and cattle exhibited at shows, had not been bred, or in some instances even reared, by the individual whose name appeared on the card. Which puts me in mind of the story, for the truth of which I can vouch, about the visitor to a farm who was on the look out for a first rate bull calf from which to breed. The owner of the place indicated one and said: "He's got the very best of pedigrees. In fact he's out of yon good cow over there". To which the would-be purchaser responded by looking the bull over and proceeding to treat the cow in the same way. Then he observed, dryly, "I can see he's out of her—but was he ever in her?"

And then, having in such ways brought their wares to the attention of the farming community, the leading breeders and all those who could afford it would send out imposing-looking mating lists. From the one I received through the thoughtfulness and generosity of the character under discussion any purchaser sufficiently green could well have gone into rhapsodies and decided that it only remained to invest enough and he would, in no time at all, find himself in the position of having bred a race of world-beaters. Which ought to have prompted some such speculation as how it came about that the whole place was not by this time alive with such wonder hens. I was also presented with a thing that looked rather like an illuminated address. This was, or purported to be, a certificate of pedigree. I cannot lay hands on it for the moment and am therefore quoting from memory; but I do not think I am very far out. I know the cockerel's dam was supposed to have run up 287 eggs in her pullet year, and his dam's dam 250 odd I think it was. Naturally all this was set out in such a way as to be calculated to make the greatest possible impact.

The notion was widespread at that period that any bird, otherwise suitable to mate up for breeding, could be relied upon to transmit to its progeny any characteristics in which it might excel. This belief, as so often happens, continued to be persisted in long after great numbers of those taking delivery of such males and females, and putting them to the test, had been mildly or greatly disappointed with the results.

Eventually Dr. H. L. Hagedoorn, the great Dutch geneticist, who had a rather special interest in fowls, succeeded in gaining some support in a number of countries for the proposition that it was the family that counted within this particular context, not the individual. In his justly famous book 'Animal Breeding' (Crosby Lockwood), he explained this by saying that, if we took two groups of hens, all in excellent health and belonging to the same genetic group and generation, but half of which had scored well over 200 eggs and the other half well under, 200 being the average, those belonging within the last-mentioned category would be highly unlikely to breed worse layers than the others.

There used, I remember well, to be a saying, 'The trend of the race is downwards'. This was subscribed to by all sorts of people, including breeders of long and varied experience. It was for this reason, and in an effort to throw everything they could into the other side of the scales,

This Dorset Red cock having such a commanding presence is mated to produce sex-linkage. The two Standard Light Sussex hens accompanying him form part of a larger group all in their fifth laying season and still performing like pullets. On one particular day half as many eggs again were collected from them as there were hens.

Photo: *British Farmer & Stockbreeder*

This good-looking bunch of Standard Light Sussex pullets went to a valued customer in Rutland, and are still there, giving sterling service.
Photo by courtesy of *British Farmer & Stockbreeder*

that so many attempted to breed from the best and most productive, even though the bird in question might have no better recommendation than that of being a diamond in a heap of coal dust.

What was true of fowls, it was claimed, applied equally, if in rather different ways, to such inhabitants of the field and farmyard as cows and sheep. And it was out of all this deliberation on the part of Hagedoorn, and his many visits to and discussions with not only breeders but also users in these islands, California and elsewhere, that what has become known as the Nucleus System grew. This represented about as big a break with tradition as we have witnessed. New-fangled schemes aimed at producing more consistently profitable stock, or doing it in better, quicker or more convenient ways—Lerner's Californian brainchild; the Cheshire System and a host of others—there have been. Some of these have remained in vogue for quite a while before being seen through or giving way to allegedly better ones; others have fizzled out rather quickly or been found impossible to make sense of by anyone other than their author. But Hagedoorn's may be said to have stood the test of time if that is what we are to infer from the fact that not by any means everybody who took it up when it first caught on has seen fit to abandon it, and other recruits have since been added to the strength. I believe this is largely to be accounted for by its peculiar nature, in that it is to some extent mechanical and goes at least part way towards cutting out the human element and rendering it foolproof. But I cannot resist quoting what some wit—I forget, now, exactly who it was—said when the *Daily Mirror*, with all its pictures, first appeared on the bookstalls: "Alfred Harmsworth invented the *Daily Mail* for those who couldn't think. He has now invented the *Daily Mirror* for those who cannot read".

In the middle of the township of Petaluma, Saloma County, California, there used to stand, and for anything I know still does, an enormous golden hen. It was meant to symbolise the principal local industry. Needless to say it was not really made of solid gold, but only got up to look like that. There are those in our midst who have set up, not a golden hen or even a golden calf, but a graven image nevertheless. A hen is a living, vibrant creature, able to respond to affection and care. But for years attempts have been made, in certain quarters that we all wot of, to ignore or deny all this and treat her as though she were a machine and therefore inanimate. A hen, like a cow, can only be studied to any sense where it belongs, in, that is, its natural setting. Yet these gentry, who are too clever by half, have now been at it for countless hen generations, having taken care first of all to place her in a totally artificial environment, attempting to discover what makes her tick and how best they can go about her exploitation, not qualitatively but only quantitatively.

One Sunday afternoon some years back, in what was alleged to be a B.B.C. farming programme, we saw a group of young ladies—charming young ladies they were, admittedly—sitting on as many stools armed each with a pad and pencil, plotting the course or studying the movements of a row of pullets in individual cages. I am not certain whether or not this was the same big official 'research station' whose Director is well known to me and who around that time felt it incumbent upon him to advise the country's poultry keepers to 'breed from the sit-downers rather than the stand-uppers, as this helps eliminate the broken or cracked egg problem'. If the birds are provided with well-lined nesting boxes, and are happy and contented, this kind of trouble does not arise. Or was it being taken for granted that at least a substantial proportion of those hanging upon these words of wisdom would be using their battery hens to produce further generations of battery hens?

Only the day before this was written a customer told me of a recent visit she had paid to a very large battery house, and she said she had been glad to get away from the place. The woman in question has long been accustomed to real hens, living more or less as nature intended. But she said the continuous, high-pitched scream which issued from all those thousands of poor creatures was the most unnatural sound she had ever had inflicted upon her ears.

And who, pray, is it that is finally landed with the bill for those capers we have just been describing and has nowhere further to pass it on to? Muggins, of course and unless he is smart enough to think of some way of getting round it.

It is highly unlikely that anyone, having read thus far without skipping any and taken it all in, remains in any doubt as to my attitude towards official testing stations or for the matter of that official anything else. But, for what it may be worth, after a number of such complexes had been set up, with the object of making an independent assessment of what was maintained or offered for sale by those calling themselves pedigree breeders, the conclusion arrived at and published abroad was that very few of them had good stock, or wares to offer that were worth anything like the prices charged.

The ensuing and rapid decline in the practice of sending out catalogues and mating lists was explained by quoting Mark Twain, who said that you couldn't fool all the people all the time. The term 'pedigree', as applied to anything having to do with poultry, began to disappear from the language until, nowadays, it is employed by few operators other than myself. And, although the label 'Think Tank' remained in embryo within the womb of time, and the act of parturition was still a long way off, here and there amongst what many fondly imagined to be the genuine upper echelons of the industry, heads began to be scratched and the grey matter stirred up in an effort to find a way out of the impasse. And so the hybrid was born. As we have already seen, birds answering to this broad description had been around in America for some while, and blueprints of a sort existed; so this move was not without precedent. But over here at least this event marked the beginning of all the talk and argument that has been going on ever since, and shows no signs of abating, on the subject of 'genetic engineering'. It was, in fact, Cyril Thornber of Mytholmroyd, Halifax, who brought out the first hybrid, known as the 101. And in mentioning his name I must give the devil his due. Cyril and I, despite our occupancy of two very different and in fact diametrically opposed camps, in what few discussions we have had over the years, have always got on extremely well together. He was one of the first people, of any standing, to recognise the value and implications of my own work as a breeder and geneticist.

I have just dug out a letter from my files which goes back many years, and the penultimate paragraph runs:

'When Cyril Thornber produced the first hybrid 101 it proved that without endless search for "nickability" over thousands of parent stocks a stable level of production could not be predicted. The five million birds sold in two years put Cyril on the map. George Mann contributed what was to be the most lucid and comprehensive text book on the subject of Hybrid Chicken and this holds good to this day'.

And now we can make our way back to those washed-out Rhode Island Reds which had been hammered for eggs even before coming into my possession and care, and that if you remember after all this time we had left being unpacked from their crate. It may be that I had been reading Buffon, Lamarck, or some writer of a like turn of mind, and decided that there might, after all, be something in the notion of the inheritance of acquired characteristics; I forget after all this time. Be that as it may I was stubborn enough, and had sufficient faith in my own abilities, to be committed to the belief that I could do with that little knot of sorry-looking birds what their breeder had not and could not have done. Now don't get me wrong—they were not utter rubbish; it was only that they looked downtrodden and as if they had never been given a chance to show what they were capable of.

It has been said over and over again that nobody can juggle with genes or whatever that were not present in some form or relationship in the original stock, any more than my bank manager, despite the fact that he is also one of my closest friends, would think much about it if I kept on wanting to swap odds and ends of money from one account to another and back again that I had neglected to put in to begin with. Yet this was precisely what I was proposing to do, or how the onlooker would view it at any rate. What I am now about to set down may look very much, at first sight, like an interpolation and as though it doesn't belong here. But, as I sit and type it, the honeysuckle is making persistent attempts at getting through the bedroom window; the whole place is literally smothered in roses; the cabbages weigh like lead and their succulence and taste cannot be conveyed in words; and there are masses and masses of flowers and vegetation neither the abundance nor subtleties of which anyone could even begin to explain in physico-chemical terms or by reference to composts and such. Those who find themselves intrigued by this revelation and who thirst after further information and guidance would do well to get hold of a book entitled *The Magic Of Findhorn* and published by the Souvenir Press, London.

But we are chiefly concerned here with poultry and how they may best and most happily be incorporated into a viable organic system. And I began by placing this little group of birds, upon which so much hope was being pinned, in a five-acre meadow that was crammed with herbs of one sort and another and lay immediately below the cow pasture on whose slopes Sir George Stapledon was to lay out his plots. Their new residence, a roomy three-quarter span building, although having a low roof, it was found necessary to set up in a deepish hole from which the stone to build the surrounding dry walls had at some time been quarried. The site lay at an altitude of 1,300ft. above sea level and the alternative would have been to fasten it down with wire cables; otherwise it would have been gone with the wind. As for the birds themselves, adapting themselves to a different and more outgoing philosophy they became more and more adventurous as time passed; and I have often seen one or another of them blown right across the field, only fetching up at the far wall. Their looks also improved and they soon provided enough eggs to be worth putting in an incubator.

One intelligent and enterprising neighbour, gone from us, alas, these many years, had waited until his firm began work on the Sydney Harbour Bridge and then called it a day. In fact he ran a flock of 250 commercial layers, going to the trouble, with the assistance of his wife, of trap-nesting them all during the six winter months, seven days a week. He took a batch of day olds, hatched from these same Rhode Island Reds, and they came into lay that September. There were no culls and they averaged 150 eggs during that trap-nesting period and he told me he was satisfied they did about the same over the next six months. which seems reasonable as it is a poor affair if they could not do as well in summer as they had in winter up there in the exposed northern hills. This, of course, adds up to an average of around 300. I know, although I have never advertised the fact, that similar feats have been brought off within recent years on the farms of at least a few customers running small flocks and taking an interest in them. But these pullets also attracted much attention as they looked a picture of health all the way through and not only had good chestnut top colour, but the requisite smoky undercolour too.

So not only had these poor washed out and put upon creatures who had been committed to my tender mercies changed in so many ways that their own mothers wouldn't have known them; what we now found we had on our hands was something that, to all intents and purposes, was greater than the sum of its constituent parts. It is, in the nature of things, impossible of demonstration; but I repeat I am absolutely convinced that, with the best will in the world, this breeder or alleged breeder up in the hills near Halifax could not have come anywhere near doing with this material what I had been able to achieve. It began to appear uncommonly as if, starting out with a sow's ear I had, nevertheless, managed to finish up with a silk purse.

When I returned to pedigree work here in the Bridport area my mind naturally went back to these strange goings on and to this incident. My late father, for instance, who held an important post with B.O.C.M. (British Oil & Cake Mills Ltd.,) and so used to hobnob with the top people both on that side and the 'advisory' side of the poultry and dairy industries, struck up a firm friendship with Bobby Boutflour, Principal of the Royal Agricultural College. He it was that coined the slogan: 'The breed is in the feed'. And now we had this globe-trotting Dutchman with his amazing command, not only of English but also of those idioms which are peculiar to this most peculiar race, and his camp followers urging that the sooner we forgot all past lessons and habits, and shifted the emphasis from the individual to the family, the better.

Doubtless each of these men had succeeded in taking hold of a few odd pieces of the total jigsaw, of a segment or two of the truth, but by no means the whole picture. It is certainly undeniable that, in the past, a great many enviable milk yields have been credited to the use of 'improved' sires when they were largely due to what came out of the bag, field and haystack.

Notwithstanding all these conflicting arguments, I was resolved that I would proceed on the hypothesis that, in the right hands, the right kind of bird, whether typical or not of the immediate group to which it belonged, could be made to pass on its most desirable attributes to its offspring. I would lose no time in taking the bull by the horns, mating up one male representative of each breed or category to one female, and see what came out of the bag.

To cut a very long story very short, each one of the various lines I now have running has at some point been channelled, like the sand in an hour glass, through a single pair of birds. What followed has fascinated and intrigued, not only myself but a whole lot of others in different places. For the benefit of any who may be eagerly waiting to swell their ranks, the way I would go about it was this. And I must explain, en route, that when pullets approach maturity, having been earmarked for my own replacements, the longer I can hold them back the better pleased I am. This is a highly skilled operation and however genuinely reluctant I might be to withhold any wrinkle or piece of information that might prove of value to others or make life easier, up to the time of going to press I have failed in all my attempts to explain to anyone, even in words of one syllable, how it is accomplished. All I can say is that I don't care if a pullet, destined for the top breeding pens, doesn't come into lay, so long as she remains in excellent health and spirits, until she is between 8 and 9 months old. So we begin, then, with a group and, having held them back to the very last possible minute, we take careful note of the one that lays the first egg. She is mated to a selected male, and from this pair flow the succeeding generations. Clearly this is to put all one's eggs in the same basket with a vengeance. And it is interesting to note that, in at least two or three instances, this first egg has appeared on March 31st.

Even then we do not proceed to broaden out all the way round, passing from sister or half-sister or brother to cousin, second cousin and so on. No breeder has ever bred more closely or kept it up for as long as I have. In fact I have never heard of one who has got very far along this road without finding the line had petered out, either because he couldn't get his eggs to hatch or because the stock he was left with had become as weak as water.

Anyway, there it is. If I knew of a closer relationship than brother/sister, or that represented by going back, over and over again, to the original hen, the founding mother, I would adopt that and keep it up year in and year out. When interviewed recently by an Editor from the *'Farmer and Stockbreeder'* I was able to point out great, bouncing birds that were the end product of a minimum of 14 or 15 generations of this kind of treatment.

A few of those picking up this book will recall that the whizz-kids who turn out this modern hybrid we have kept on running into as it progressed claim it is the result of the bringing together of two or more highly-inbred strains or breeds. Only those offspring who stand up to a series of rigorous tests are retained to carry on the race. The rest are ruthlessly scrapped and the wastage is just nobody's business. The proportion of those falling by the wayside is very high. I have no wastage. I do not even have any culls, and neither do any of my customers. There is no such thing as going round handling and inspecting point of lay pullets and throwing out a whole lot of them.

On a very much milder level, important laying trials have been won by men, some of them known to me, who after maintaining two separate lines for years, have brought them together for this special purpose. But that, as Churchill said about the war, can only be done once; or at least the breeder cannot feed back the progeny if the lines are not to be allowed to run into one another.

Of course one of the objects of inbreeding is the bringing about of greater and greater uniformity in the resulting stock. But too often it turns out that the breeder, while plugging away at this, only succeeds in tipping out the baby along with the bath water. He is like the goalkeeper who, having run himself out of breath carrying the war to the enemy's camp, finds it impossible to get back to his post in time to prevent half a dozen goals being kicked through the door he has left wide open. However, as I once pointed out in an article, to import a sire from even the most apparently reliable source and introduce it without any preliminary testing, a somewhat lengthy process, to one's own stock is a very risky proceeding indeed; and I have known established breeders brought to the verge of ruin in this very way. The operator who knows what he is up to is far safer in keeping his flock closed even if it is a comparatively small one.

The idea behind the recombination of the separate lines within which this inbreeding policy has been pursued is to recover a few of those prized marbles that have on the way fallen out of holes in the bag. What is in any event gained is an access of heterosis or hybrid vigour, 'hybrid' being a term first used in 1866 by Mendel, the patient, plodding Austrian monk who chain-smoked cigars as some do cigarettes.

However convenient some may find it to pretend that this is not so, inheritance or the bringing about of any increase in the numbers of poultry, or indeed of any other category of farm stock, bristles with imponderables. And nobody really understands how heterosis comes about. Darwin had his version. Hagedoorn made a brave though unconvincing effort to explain it by reference to a pair of curtains each fairly liberally besprinkled with holes that, whenever the two were hung in front of one another in various positions, corresponded or failed to correspond in different ways.

Whatever may be the real explanation, from a purely commercial point of view, when eggs or table chicken are the object a first cross, all other things considered and being as they ought, is a better bet than a pure breed.

I should like to end this chapter with a few words about the Dorset. When I made this breed, in 1970, I used nothing but pure Rhode Island Reds on both male and female side. In fact the genetic break I was able to engineer, and from which this new and distinctive breed arose, followed the mating together of a small group of full sisters, all mature pullets, and their full brother. What is known as a sib mating. We are concerned with two separate hatches. The first came off on August 1st and the second on the 17th. I have the exact figures in my office; but for our present purposes it is sufficient to know that both Dorset Reds and Dorset Whites were produced each time, and thereafter the parents reverted to the throwing of progeny like themselves.

It was quite evident from the start that there was something different about these chicks, but this was not easy to define and, apart from 'liveability', which only answered part of the question, I found myself unable to put a name to it. They were certainly special—there was no getting away from that—and vigorous. The Whites looked much like the young of pure Light Sussex, but the Reds, so far as down colour was concerned, were all buff and impossible to distinguish from the sex-linked pullets resulting from a cross between a Rhode Island Red male and one or more Light Sussex females.

On reaching maturity the white ones came to bear the markings associated with pure Light Sussex; only most of the pullets sported a few spots of gold and the cockerels wore a light golden saddle. Their present-day descendants, although heavier, are so much like Standard Light Sussex in appearance that they are often mistaken for them. As for the Reds, the males became darker as time went on; but there was no mistaking the fact that they were of a distinctive breed, as for one thing they were racier than any R.I.R. cockerel and far more prone to attack interfering humans. And the females retained their buff plumage and looked much more like sex-linked R.I.R. X L.S. than pure Rhodes. Also the uniformity amongst these Reds was, and still is, astonishing. They come like peas out of a pod.

It must be pointed out that, although Dorset Whites and Dorset Reds are, in many respects, quite different animals, like white and red Shorthorn cattle they both belong to the same breed, and one can produce the other; so that if for any reason we were to run out of either this deficiency could soon be remedied. For when Whites are mated together we end up with mostly Whites but also a proportion of Reds. These last are described by my friends at Stanbrook Abbey as honey-coloured, but when grown up they turn into authentic Reds—except that, up to now, all turn out to be pullets. In my own test matings sometimes we have had all pullets, and at other times one in ten or twelve has proved to be male. Nothing of the sort has ever been encountered before in connection with poultry anywhere in the world. In fact I do not think it would be any exaggeration to say that the Dorset adds up to just about the biggest paradox ever to have hit the world of biology or genetics.

To tick off all the various other relationships and combinations which have been tried out over the years:

D.W. male on D.R. female gives us about half authentic Reds and Whites, male and female; Red on Red produces about two-thirds Red and one third White, male and female; and D.R. X D.W. results in just short of 100 per cent sex-linkage.

Without having any desire to confuse anyone, or blind them with science, this is the name we give to criss-cross inheritance where, because one chromosome is present, along with what we might term a blank cartridge, in the female ova, but one in each of the male sperms, and dominance comes into it, the pullets come gold like their mother, and the cockerels silver.

However, when Mr. G. M. Nightingale, of 88 Elwell Road, Upwey, Weymouth, obtained from me a pure Dorset White cockerel and mated him with a group of pure Standard Light Sussex, also obtained from me, he reported that all the pullets came gold, with the males silver.

Now, supposing we were to turn the thing round and put a silver male, let us say a Light Sussex, in with a golden female, say a Rhode Island Red or Buff Rock, because the factor for silver is carried on the sex chromosome, every chick hatched would have silver plumage. Only although they have the outward appearance or phenotype of pure Sussex, they are, in fact, heterozygotes. Which means they are impure for the factor we are here dealing with. And, whilst we are at it, we may just as well go a little further and explain that, while a bird that looks like a recessive cannot be anything other than a recessive, one can be indistinguishable from a dominant without being one. In plainer and simpler language, in certain circumstances a bird can look exactly like a pure breed and yet be a first cross. And the only way of finding out, if we are not sure, is by undertaking a special kind of test mating.

It will, I hope, be clear from all this that, when dealing with such things as gold recessives and silver dominants, on the basis of a few simple mathematical formulae it is child's play to work out beforehand the various ratios we are going to end up with. Or at least it was until the Dorset came along. This breed, as we have already seen, has upset the Mendelian apple cart good and proper. The veriest layman or tyro can see from the foregoing that dominance is all over the place, the same individual behaving exactly like a dominant in one relationship and a recessive in another.

This is why so many graybeards and eggheads in different parts of the world—professors of this and that, and those engaged in various fields of research, real or alleged—when handed anything in which my name or that of my brainchild figures, drop it like a hot potato. There are still plenty of those around who cannot bear to have their cherished illusions shattered or the idea that, like those holy and august men who refused to look through Galileo's new-fangled telescope, they might see something which, if they were not to live out the rest of their lives a lie, would cause them to cast all the books, stuffed as they are full as a plum cake with dogmas and articles of faith, into the fire, and tear up the whole of the sacred curricula.

No child relishes the prospect of having to give up his teddy bear in exchange for something a little more grown up.

The trouble, truth to tell, with every orthodox scientist that breathes, no matter what his discipline, is that in the nature of things and by his very definition he is completely debarred from studying to any sense, or taking up the kind of attitudes towards any piece of phenomena you like to fasten upon which alone can render a valid conclusion possible, from where he is standing. And, being the sort of animal he is, there is no hope of getting him to shift his ground. Great and worthwhile natural secrets never show the least inclination to tumble into the lap of the sceptic, the uncommitted, the scornful or the irreverent.

Open Day at Stoke Mandeville (B.O.C.M.), 1956. The pullets occupying this traditional henyard are believed to be Thornber 101.

Photo by courtesy of *Poultry World*

This is how the interior of Dr. Sainsbury's original henyard at Cambridge looked shortly after installation by the suppliers, Andover Timber Co. Ltd.

CHAPTER FIVE

MAKING THE BEST USE OF LAND AND HOW MY RANGE/YARD SYSTEM CAN HELP

When, in 1884 Disraeli, believing himself to be doing a wonderful and humane thing, for which posterity would applaud him, extended the franchise to embrace the agricultural worker and those occupying the more humble berths on the land, he could not well have done them and the rest of us a greater disservice. In handing them the vote, far from extending their freedom, he diminished it. He ensured that they remained enslaved—barring something in the nature of a miracle—for all time.

Since that fateful moment in our history, not only have those filling the lower echelons in the rural hierarchy handed over their dignity, independence, and the bulk of their responsibilities to others; the canker has managed to creep all the way up to the top. As a result many stately homes have crumbled because of the astronomic cost of maintaining them in repair, staffing them and so on; or else they have been replaced by monstrosities of one kind and another put, as likely as not, to an equally monstrous use.

A few of our long-established landowners, small as well as large, have managed to survive in a fashion by trimming their sails and resorting to all manner of devices none of which need ever have been necessary. With the arrival of that iniquitous measure, Capital Gains Tax (Circa 1966), a big stone has been hoisted on top of that already existing one that used to be called Death Duty but has since been promoted to Estate Duty, and more recently still to Capital Transfer Tax. I expect it was decided that such a title would sound less morbid and more respectable.

What with these and a whole range of other impositions, restrictions, edicts and embargoes, many estates have folded up. The Compulsory Purchase Order has placed an exceedingly useful tool in the hands of the vandals. The most impudent and high-handed instances of highway—or motorway—robbery have been reported. And plenty of little people have suffered, too, quite in addition to those who and whose parents and grandparents had been accustomed to live in the shadow of the great house and whose fortunes were therefore bound up with those of 'the gentry', and who had believed themselves to be part of an institution which, still largely feudal in character and not a bit worse for that, they had never thought of as vulnerable or ever coming to an end. Magna Carta has been thrown, contemptuously, out of the window. And simple, honest folk in great numbers, who had believed their hearths and homes to be safe and sacred, protected as they were by ancient and inviolable covenants, and asking little of life beyond the right to be let alone, have been interfered with in the most gross and brutal ways. They have been literally torn from their homes and community roots and sacrificed to that sacred—or perhaps we should have said profane—crocodile which possesses an insatiable appetite for land and a weakness for outsize Meccano sets with which to construct such eyesores as the one occupying a site at Sutton Coldfield, amid the once fair Warwickshire countryside.

But let us not, even while saying all this, err on the side of complacency. For often things go by default, besides resulting from some calculated act of wickedness or from cupidity. The situation is already desperate, though not beyond hope of remedy, and there is no room for apathy or lack of vigilance. Each fall on Long Island, New York, a somewhat curious phenomenon takes place, and various strange rituals are played out by a large section of the community. For instance, with one hand it collects up dry leaves into heaps while with the other it harvests or buys or offers for sale great numbers of swollen pumpkins, to be made into pumpkin pie. I have not heard that any of their number ever see any inconsistency in this, or have any inkling that what they are helping to do, in tearing off a match to put to the heap, is to destroy their own heritage and prepare a funeral pyre for their descendants. Pumpkins will no longer be able to wax fat, or even develop at all, once we have destroyed the wherewithal to provide their sustenance. Day after day, this side of Southampton Water, that side of Brooklyn Bridge and in fact all over the place, great mountains of prime and irreplaceable organic matter are wantonly or thoughtlessly destroyed or rendered useless or inaccessible, in a world that is tragically short of humus. Just as I can never make out how so many members of *homo sapiens* can cheerfully churn out great quantities of tenth-rate poultry meat or eggs all week and be looked up to and respected as active members if not pillars of the Anglican or some other Church on Sunday, so my curiosity is piqued over and over again as I encounter and hold converse with those who are mortally afraid lest they should be caught in even the most minor offence against the Holy Ghost—for neither the precise nature nor indeed the very existence of which or whom no shred of evidence has ever been adduced that would stand up for five minutes—yet are innocent of qualms of conscience when it comes to deciding upon the most holy or even sane and sensible method of disposing of and putting to use the wastes from their own bodies. Despite the never-ceasing endeavours of watchdogs and campaigners like myself, they are able to contemplate with equanimity the notion, if not the actual sight, of huge quantities of potentially valuable sewage being dumped in the sea.

All this lies very close to the hearts of every one of those men and women who qualify to have the adjective 'organic', within our present terms of reference, attached to them. Day after day people from every point of the compass, but all with much in common, sit opposite me in this room, and out it all comes. They express their concern about what is happening to the quality of life and about the creeping canker of materialism. They want to know what they can do to enlarge human freedom and get the incubus off the backs of those who could make such an enormous and significant contribution to the larder, health and well-being of the nation if allowed to do so.

In particular are they concerned over the scarcity of suitable holdings and the fact that many of those who, by tradition and heredity, belong in the fields, the cowshed or the stable are unable to find places to rent or otherwise to gain a foothold in the countryside.

When every anomaly and abuse has been duly catalogued and the worst has been said that can be said about the real old-fashioned landlord, it has to be conceded that he rendered an important and, in fact, essential service. Nor has anyone come within miles of finding anything worth calling a substitute for him.

All those responsible and concerned folk with whom I have spoken are agreed that we must take arms against the ones who have just been listed. We must equip ourselves with the most effective weapons we can think of or lay hands on, short of billhooks, pitchforks and the kind of instruments those French peasants and what not hastily shouldered before making a bee line for the barricades and the Bastille. Somehow that army of parasites, some of them smug-faced, others completely faceless, must be driven out or as an alternative rehabilitated and found some kind of honest work, that they may in time and no doubt for the first time gain self-respect and become clean and wholesome. Meanwhile we have got to make what shift we can and be content with such land as we are able to come by. What we can at least do, and what is surely incumbent upon us, is to make the best possible use of what we have.

It is, for instance, an irresponsible treatment of land and other resources to set out planless and, dumping the new or second hand huts off the delivery lorry or the one that has followed the furniture van down on the nearest or most convenient bit of more or less level and more or less clear ground to hand, to become caught up in a hundred other tasks and enthusiasms so that they get left there, henless or with ill-thought-out stocking, for the next year or two. The other mistake so many beginners make is to site the houses, coops and, in fact, the whole bag of tricks unwisely; the result being that they end up linked to each other by a series of narrow trackways beaten through a jungle of nettles, brambles and other uneatable growth and with nowhere for the inmates to graze or even move about freely. Nettles and docks are extremely valuable in their proper place, making as they do their contribution towards putting the manufacturers of, as well as traffickers in, chemical fertilisers out of business; and copses may be incorporated into any attractive and workable scheme. But we cannot afford to treat so wantonly a rapidly diminishing asset, and fowls must be provided with the greatest area of palatable sward that can be managed.

The reverse of this, but just as unacceptable, is provided by the sight, not all that uncommon in different parts of the country, of a higgledy-piggledy group of folds and other movable appliances that never get moved from one year's end to the next. The result is that one small area of the bailiwick quickly becomes completely bald and stays like that, while grasses flourish unchecked all the way out to the boundary fence.

Now the author of these various observations would be the last to deny that any organic unit or entity must go on evolving if it is to continue to be so defined, if it is, that is, to remain healthy and make its optimum contribution to the common weal. The stream that is cut off at both ends turns into a stagnant pond, or else dries up and in either case is of little use or interest to anyone save, maybe, a pathologist of sorts. And with all my agricultural and horticultural experience and background, when I first bought this immediate property I did not feel it right or natural that I should sit down and draw up a ground plan: a scheme that would take in every corner of the garden and grounds and to which they would be committed for all time or as much of it as we need concern ourselves with. A famous firm, situated not more than a thousand miles from Ascot, once tried hard to persuade me to take up a post in which I should have had to supervise the work of a large number of landscape gardeners operating in different counties. But I showed no enthusiasm. In fact I can think of several areas that it took years of just standing and gazing at, and brooding over, and discussion on the part of my wife and myself to legislate for to any sense or with any hope of providing long-term satisfaction. Just the same, when laying out cash and committing one's future to a poultry enterprise which it is intended shall act as the king post in the whole structure and around which the Jerseys or Dexters, the sheep and the goats and all the rest shall revolve and whose ends they must subserve, it is rather important to start, so far as lies within one's power, as one means to go on.

It is all the more necessary to stress all this in view of the fact that no single aspect or department, within the whole orbit of poultry husbandry, has succeeded in generating more trouble, frustration and loss both of temper and cash than that dealing with the housing and disposition around the place of laying stock, particularly laying stock. Such things as breeding pens, unless flock-mating is being practised and the females are therefore being run in large numbers, are much easier to cater for, as the nature of their habitat is determined by the fact that we are likely to be dealing, in each case, with a unit of ten to fifteen birds which must, the nature of cocks and cockerels being what it is and some hens possessing as great an appetite for extra-marital adventures as some women, be securely fenced in or, as in the case of some of our own free range breeding pens, separated from the nearest source of temptation by not less than half a mile of fields and hedges. But, when it comes to commercial layers, pure and simple, just about everything that human ingenuity could contrive or that desperate, half-crazed men and women could think of or rake together the materials for has been tried. And it was because most of these had failed dismally and none of the rest had ever proved 100 per cent satisfactory that I came up with the system which, in use here and there in various counties, has at last put an end to all the stopping and starting and made it possible for the previously encountered snags to be overcome and the poultryman to sleep in his bed o' nights unaccompanied by spectres and nightmares.

This method, which as I say promises to put an end to the so familiar sight of fields and odd corners littered with the evidence of past mistakes and failures, has been christened the *Thompson Range/Yard System*. That label was attached to it, not by myself but by a valued friend of mine, the distinguished authority Dr. David Sainsbury of the University of Cambridge, in an interview with the Press. The reporter at the other end

The author found that he had not only bought the ground underfoot, but a great weight or organic matter above it. This, however picturesque, did not provide anywhere for fowls to roam and graze, or for crops to be raised.

So there was only one thing for it. He put the bulldozer through it and followed up with a forage harvester so as to chop the smaller and softer stuff up even smaller, to be later pushed into a heap for compost.

of the line or occupying the chair opposite was another David, David Wilson of Weymouth who has the distinction of being the first newspaperman anywhere in the world to have instituted, quite off his own bat and with nothing much to go on, a ferreting-out process in connection with the history-making Dorset breed. He has since become a good friend and I am very glad of this opportunity to publicise this example of intelligence and enterprise. As for David Sainsbury himself, whatever he may have to say is carefully listened to in whatever country he may chance to be sojourning at the time, and his dignified, restrained and responsible approach to all matters pertaining to livestock has commanded much admiration and respect. Although, as already indicated, this has been strictly a one man show from the beginning almost to the end, I have enjoyed his support and encouragement all the way through. Out of the whole academic and scientific world he was the first to come forward and say that he recognised the significance of the work I had been able to do here. When, therefore, he unhesitatingly accepted my invitation to supply the Foreword to this bundle of heresies and, for good measure, expressed his delight, I could only reply that he would be hard put to it to be more delighted than I was.

Having thus landed myself with the task of explaining what all this is about and translating what could well be a highly technical and obscure discussion into layman's language, I had better get on with it. Now just as, in a 12-bore, we few authentic and old-fashioned farmers who still remain use that size of shot which is best adapted to the type of quarry we are after (e.g. no. 7 for snipe; no. 5 if it is rabbits, wild mink or wood pigeons we want to bring down; no. 1 in the case of the larger, tougher or thicker-skinned predators and vermin such as foxes, V.A.T. or income tax merchants, 'the men from the ministry' and so on), when it comes to fowls whatever we are doing or installing must be geared to the nature and size of the enterprise and the number of birds for which we are setting out to cater.

There is a certain place in Suffolk, associated for all time with the name of the late Omar Kháyyam, which I confess I have seen very little of in daylight. I did indeed land there once, in a Lancaster bomber, and spend the rest of the night very happily and comfortably, drinking coffee made with the maple syrup I am so fond of, and one thing and another, as a guest of the Americans. But more recently one of its residents evinced a desire to put up a range of these yards on his farm, the first of these making its appearance in the orchard. And, before going on to what may sound like a criticism, I had better soften him up by saying that he has made a jolly good and workmanlike job of interpreting my drawings. Now this model, the MK II, is intended for occupation by fifty adult birds, and my customer and correspondent has, up to now, only succeeded in mustering a total of four. In fact when I heard about it I became so incensed that I immediately drove at breakneck speed to the station and sent off a further fifty day olds. Naturally I could not charge for these, but the recipient paid for a second batch of the same size. So he should end up with a minimum of 50 pullets, plus the original four, if these have not in the meantime succumbed to loneliness or got lost in a corner.

But once again we must retrace our steps along the road we have come, or I have come, to let out a few tucks that were necessary to a shapely garment. We must, in fact, go back to the time when I first began to ponder the possibilities inherent in what had become known as the henyard or strawyard system. Geoffrey Sykes of Salisbury had, rightly or wrongly, been given credit for its promulgation as any kind of a formulated scheme, which could be conveyed in words, drawings and photographs; and in fact he wrote a book on the subject. There have been many versions of the henyard; and most of these variations, where not dictated by the notions of the farmer concerned as to what a henyard should look like and how it ought to function, have come about through having to be tailored to fit a particular climate, situation or pocket. For instance, the shape, capacity and layout of existing buildings which happened to be going begging and so could be pressed into service; the availability and price of straw: these were only two of the many factors which, at one time and another and in one set of conditions and another, popped up to determine the final pattern. Geoffrey and a great many of his contemporaries had the advantage of plenty of cheap straw. I myself have had lorry loads of excellent straw almost thrown at me. A few years later it was to become like gold.

A conventional henyard must be constructed on land that drains well, but at the same time if it occupies anything approaching a steep slope all the litter tends to end up at the bottom and to have to be carried back. All sorts of dodges have been resorted to, in different places, in an attempt to overcome this difficulty, the favourite being the positioning of a row or two of bales of the same material athwart the incline. The idea behind this distinctive method of housing layers is, of course, that the inmates, secure from foxes and sheltered from the wind, may nevertheless enjoy the fresh air while scratching in straw for corn, and in some cases that I know of pellets. What caused it to begin to lose supporters, even before there began to be difficulties with the supply of straw, was the incidence of coccidiosis, intestinal worms and suchlike. One 'expert' described the average strawyard to me as a glorified muck heap.

It did, however, occur to me that, if these various snags could be overcome or if, better still, it could be established that they need not arise in the first place, the maintenance of flocks under such conditions offered many advantages denied to those who in any case had nothing much else to fall back on. After all one had, on far more occasions than could be kept track of, watched brave new ventures come into being, borne up on the energy and enthusiasm of those who had set out to do better than any of their predecessors. One had seen houses bought and installed and runs put up, and the new acquisitions fondly and tenderly placed inside. And one had lived to witness the gradual draining away of the adrenalin and the inexorable wearing down to a threadbare state of the carpet of turf—when there was ever any present to begin with—and the puddling up or souring of what remained underneath. It is perhaps, and ironically, one of the biggest bugbears attaching to such ventures that the whole setting can look so fresh and pristine before the first egg—if indeed there ever are any eggs—comes to be deposited in the smart new nesting box, and to give no hint of the holocaust to come.

One had, in one's extreme youth, been present, and as like as not given a strong and willing, if unwitting hand, when smallish field houses, perfectly good in themselves, sensibly and stoutly made by local joiners and suitable enough for the accommodation of birds in that latitude, altitude and climate while the sun shone and the gentle zephyrs played over the hillsides, had been carted and strung out across a series of pastures. And one had watched the pages of the big Silcock's or Bibby's calendar that hung on the wall of the kitchen cheek by jowl with the sides of bacon being turned and the months slip by—time that can be such an enemy when everything has to be kept up to scratch if egg production is not to fall and all sorts of troubles and vexations develop, and the daily round must be maintained and distances traversed and things carried to and fro, with none of these chores growing any lighter or less irksome with the falling away of the first fine, careless rapture. And of course, when the really bad weather came along, and particularly as the first snowflakes began to drift down towards the homestead and surrounding fields that really, in most cases, put the tin hat on it.

Even when, as in some cases that are still clear in my memory after more than forty years, it was decided to put up one or two large, substantial cabins out in the fields so that the birds could be kept inside during the kind of weather we have just been discussing, and the surrounding area looked spick and span and everything went more or less like a dream, according to who it was that stood at the wheel, there were always and inevitably squalls of one sort or another ahead. For a big house, without wheels or skids and nailed up good and solid, is not intended to be moved on every now and then, and cannot be. It has to stay put, in many cases, for more than one owner's lifetime, becoming part of the scenery. And according to what lies underneath, the surrounding area of grass becomes slowly or quickly poached and often fowl sick into the bargain; and nobody I have come across in such circumstances ever knew how to prevent it from staying that way and in fact getting worse.

What I have just been describing will be recognised by the more enlightened or two or three times bitten as a kind of semi-intensive system. The advantage claimed for such systems is that the occupants can be kept in when conditions outside are not conducive to comfort, or only released for a few hours each day at any time in the hope of keeping the surrounding apron of ground from becoming foul. In some cases alternate runs were provided, so that at intervals one pop hole could be closed and all the land running out from it wired off in order to give it a chance to recover, and another opened. Whenever this was done the high wire netting fences that were necessary, once erected, stayed up. For dividing wire that has to be taken down and put up again somewhere else at frequent intervals can be the cause of a great deal of trouble and, on occasion, blood-letting, and is in any event very time-consuming. Systems like the Borden, which call for the provision of a series of very large and heavy hurdles, to be moved from place to place as required and set up much as we used to when folding sheep over turnips or rape, tend to be very cumbersome.

The way the late George Henderson circumvented a few at least of the drawbacks normally encountered by those who run poultry in such a manner was by means of a species of compromise. Like all the rest he put up rows of detached and well spaced out cabins in his grass fields but, turning the universally applied formula inside out, he let the birds run out through all the worst weather, only shutting them up day and night when the best came along. This was so that they did not tread down the long grass until it had been cut for hay. It was an attempt to have the best of both worlds.

For over a decade now I have run continuous trials here on the strawyard system as originally conceived. Parallel with but quite independently of these Dr. Sainsbury at Cambridge has been doing the same. I have not seen his set up, nor have we ever compared notes. All he has ever said is that his results and conclusions are very similar to mine. I shall, therefore, be as interested as anyone else to read what he has to say when he comes to write it up.

There have been two chief motives behind my decision to set up and continue with these experiments. To be sure, from what the theologians call an exegetical point of view, the title on the outside of this work being what it is, I must place the emphasis squarely upon the keeping of feathered stock outdoors and if at all possible on grass. And I am going to affirm, as I have done on so many previous occasions, that where suitable facilities for free range exist, there is no point whatsoever, on economic or other grounds, for the pursuing of any other system with pullets or hens kept simply to lay eggs for human consumption. We are, none the less, confronted with a whole range of physical and other circumstances which, as we have already observed, militate against any likelihood of all those who would dearly love to get their hands and feet on some land being able, within the forseeable future, to come by enough or in some cases any at all. The needs of these must be catered for, and we have a duty to extend to them all the help and advice of which we are capable. My wife, during one phase of her career as a schoolmistress and already a keen and competent gardener who could persuade just about anything to grow, had to make do with a long row of window boxes. She has been luckier than some.

To such I can only address the following: Given stock with the right kind of background and rearing, eggs of reasonable quality—I am not going to tie my hands by putting the case more strongly than that—can be turned out in economic quantities by the use of strawyards. This assumes that the sun is going to be able to get at these birds and constant raw greenfood, possibly combined, in season, with zero grazing, which would often mean the use of someone else's mowings, is not only available but available fresh. I myself have pullets laying very heavily in such conditions. But they would lay better and for longer, and be a much better proposition financially, out on range under my system. In fact

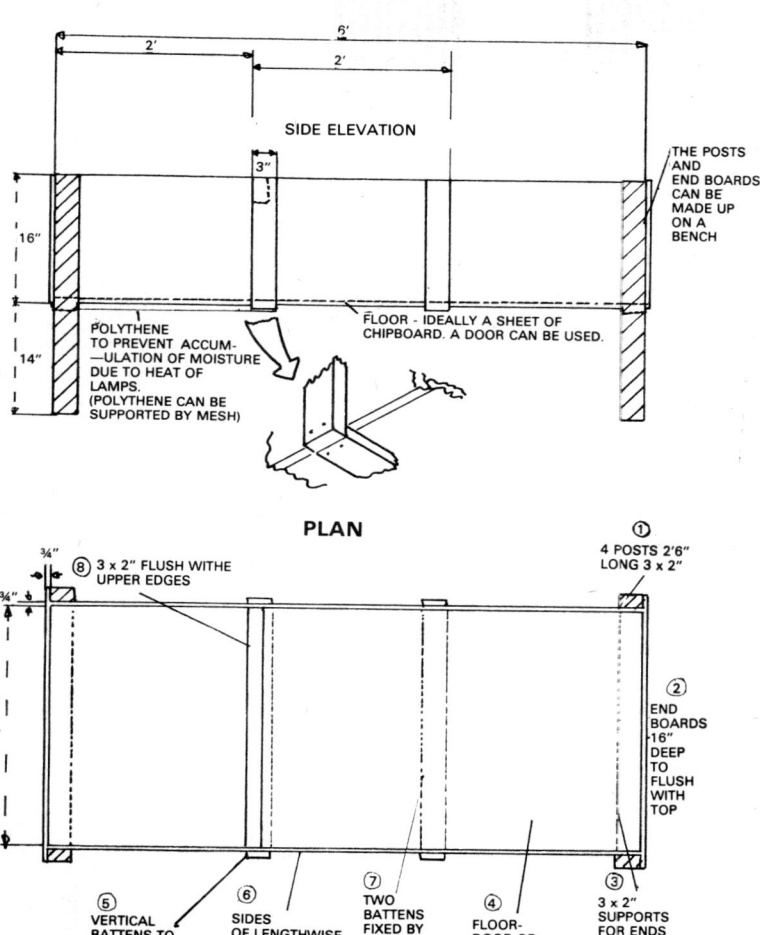

160 tons of valuable compost every year. In both of these pictures the pit is shown empty so that the exposed pipes and tiles can be seen which provide just the right amount of **ventilation**.

these are Dorsets, and the higher and more exposed the situation, the better this breed seems to do.

I think that represents a reasonable and sufficiently adequate statement of the position.

In mentioning all of which we have involved ourselves, as so often along this particular trail, in a lengthy explanation, this time of the most suitable kinds of greenfood to go in for in these special circumstances. But, before plunging, knee-deep, into any such growth, something had better be said about those marginal situations where the householder may indulge in wishful thinking to the extent of using what little ground he (or she if it is a woman) is able to set aside for such a purpose as a place for the hens to run out on, instead of putting up a yard suited to that number of birds and setting their manure (usually reckoned at about one cwt. per adult bird per annum) to work raising a succession of crops for their enjoyment and sustenance.

Having reached which point let me issue a word of warning. For whereas the hen in her infinite wisdom, or as a partaker of the infinite wisdom of the universe and of nature, when free to do so applies her droppings in situ and spreads them over an area, far too many humans, ignorant, indifferent or unaware that material of this kind must be used judiciously, dump great dollops of it down wherever they can find a space, or pile it in great, soggy heaps. If it has to be stored by itself, it must be dried off and placed under cover. This is commonly done by using superphosphate. But superphosphate is made by a synthetic process, whereas such things as rock phosphate or carbonate of lime: these are natural products.

I never have the problem of storing poultry manure on its own. My way is to compost it, and for this purpose I use carbonate of lime. Over a period I never made less than 160 tons of compost in the course of a year; and 100 tons is nothing.

Now as to the sort and range of crops that are going to lend themselves best to such a scheme. One's thoughts naturally jump in the direction of some members of the brassica family. And we shall certainly grow some of these, whatever else we may, on further reflection and advice, add to the catering list. The kales are more valuable in many ways, because of their vitamin content and so on, particularly for breeding stock. But cabbages are so handy and it always seems impossible to get by without finding accommodation for at least one variety. Yet cabbages, gross feeders and eaters up of nitrogen though they are, reach a point where they cannot touch another bite; and with all that beautiful manure still to find a home for, and a few spare corners of land that we missed at the first count, what better than Russian comfrey? You can't over-manure that, unless you bury it so deep that it disappears from human sight and knowledge.

Just as the 'Small Farmer Scheme' began as an idea in the fertile mind of Jesse Collings, who having been born at Lympstone in Devon, afterwards moved to Birmingham and in 1886 made his famous demand for "three acres and a cow" for the agricultural worker, so in 1870 Henry

Doubleday, a Quaker from Coggleshall, introduced Russian comfrey as an answer to human hunger. And now I can find the opportunity, for which I have been keeping a look out all the way through, to pay a warm and much overdue tribute to a staunch and loyal friend and colleague of many years' standing. I refer to Lawrence D. Hills, founder of the Henry Doubleday Research Association in 1955 and its secretary ever since. If there exists a greater, more dedicated or more selfless horticulturalist in the world today I should be interested to meet him. It used to be a joke amongst his friends and supporters that Lawrence Hills 'made his living in his spare time' in order to devote so much labour to his work and this cause. And people have come up to me, at Chelsea Show and elsewhere, and said: "That man should be given the highest decoration there is going and is appropriate".

I do not, personally, go much on such things; but I have often had cause to ponder the direction our priorities have taken over the past decade or two, when one reflects upon the fact that one man can go hard at it for several years, reducing what was once a good, if monopolistic, postal service to ruins and be ennobled for his pains, while a second can have the same reward bestowed upon him for only just falling short of bringing the country to its knees.

The only reason comfrey is not better known than it is, and its properties more widely recognised, is that nobody has yet been able to think of a way of making a corner in it, or turning it into a money-making racket. The fact that it might be of use in filling a few million empty bellies is of little consequence. Rich in protein and several of the more important vitamins, this plant is capable of producing bumper crops according to the climate and soils when properly managed. In fact it comes as near to realising the age-long dream of perpetual motion, when existing in symbiosis with poultry, as anything I can think of. It is, admittedly, somewhat difficult to harvest and convert into a state in which it may conveniently be fed; but great strides have been made in this area.

It was around 1960, when I owned a farm in the New Forest on which I ran 2,000 head of stock, that Mr. Hills asked me if I would do some work on poultry manure and Russian comfrey. I happened to have an acre and a half that I could get ready and devote to such a purpose, so I was glad to do so, for the value of poultry manure was not and still is not sufficiently appreciated. The rest of the story, and any further information about all this may be obtained from the H.D.R.A.

Very well, then, let us, as this book goes into its last lap, get down to the business of explaining what precisely this range/yard system consists of that has finally emerged from all this research and these many trials and all the stirrings of grey matter.

It will be seen from the illustrations that this modified henyard is roofed over but that there is a strip running all the way along the front which, except for wire netting, is open to the sky. The whole structure should be carefully sited, with its back to the prevailing wind, if it proves impossible or impracticable to find a spot so well provided with natural shelter that the wind makes no odds. It will be noticed that the front

"BRADSTOCK BROODER" — CONSTRUCTION
PARTITION AND HEATED COMPARTMENT

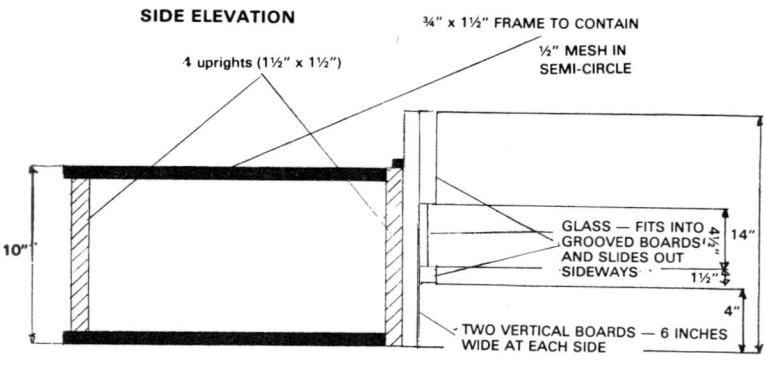

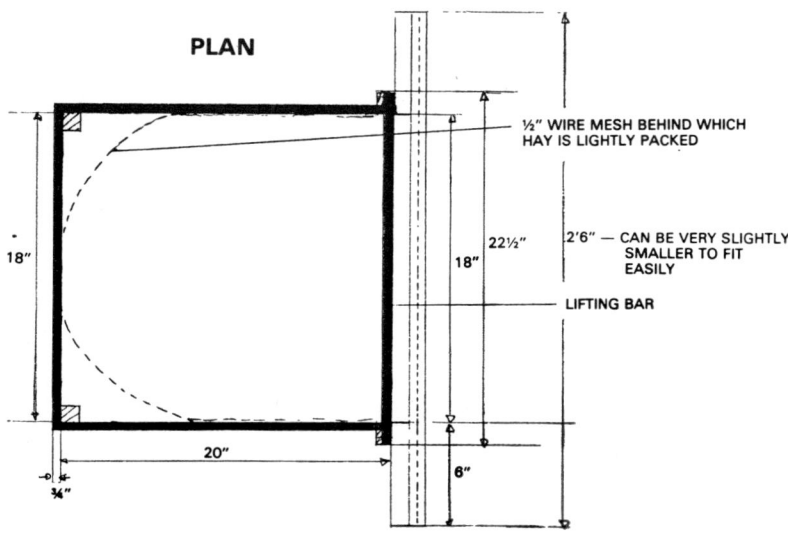

BRADSTOCK BROODER — CONSTRUCTION
ASSEMBLY: LID AND LAMP CAGE - DETAIL

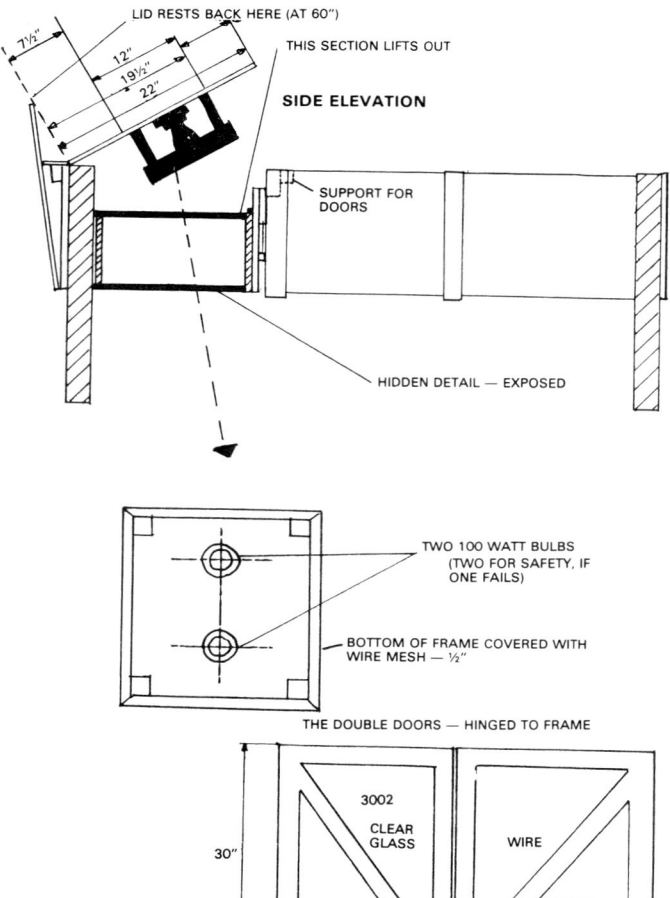

and sides are open above a certain point. A fowl's respiration rate is quite different from that of a human being and she needs all the fresh air that can be provided. A house 21½ft. square will give fifty birds one square yard apiece. In this way the attendant has a good deal more leeway than with any other kind of housing. Not only can he shut them in, or allow them to stay inside if they want to, during bad weather; he need not let them out in the mornings until he has finished milking or considers them safe from foxes and other marauders, but can rest assured that they are safe and happily occupied in scratching amongst straw for grain.

A word about the litter. This consists of plain, straightforward straw, and is cleaned out at regular intervals. Deep-litter conditions must in no circumstances be allowed to prevail. I make a special point of drawing attention to this since even some of my friends in different parts of the country have, at one time and another, expressed their preference for this Transatlantic import in certain circumstances. Deep litter is a totally unacceptable and downright unhealthy way of keeping chickens and is not even a necessary evil. Anybody with a pair of eyes and ears as well as a nose should never be left in any doubt on this score, having stuck them all inside the door for a few brief moments.

Oh—and I almost forgot—what we do is to put up a duplicate set of these: one on one part of the farm and the other as far away as the total acreage and the overall plan and system allows. One set is in continuous use for exactly a twelvemonth. Then the whole flock is moved bodily to the second. In this way none of the land ever becomes poached, but is rested; and there is much less danger of disease or infestation.

All of which naturally raises the question of cost. I could dig out the figures relating to the range of yards which I put up, some years back, to accommodate a total of a few hundred birds in units of 50. But they would be hopelessly out of date and misleading. For instance, in my costings I put down labour at ten shillings per hour. This would no longer be anywhere near realistic. Conditions vary from place to place and from operator to operator. All one can say is that, if care is taken and the cheapest possible materials used, consistent with suitability and durability, the cost per hen-housed bird should compare very favourably indeed with that obtaining under any other system. Slabbing, offcuts, second-hand flooring, matching and joists: in most localities there should be enough of these to be picked up by those who keep a sharp look out and have a tongue in their heads.

It will be seen that the perching arrangements are such that, not only are the residents well down out of the wind, where there is any, but they can please themselves whether they roost under the roof or the sky. We have already found that longevity in the laying hen goes hand in hand with the provision of whole grain. The enormous importance is now going to be stressed of seeing to it that, if she displays a preference for being rained, or even snowed on, all night long if need be, every encouragement and facility is given her to do so. Some of the best hens I have ever bred could not have been induced to spend the night under a roof even if there had been those around with any such ideas in their heads, but would insist upon sitting in a row on top of the ridge. At least half of my homing pigeons are the same way inclined; and the harder it rains, the more they seem to enjoy it.

I have heard it said that a man can sometimes live with a woman for a great many years and at the end of it know next to nothing about her. And it is strange, but no less true that, whether because he is blind as a bat or is keeping them under conditions in which they have no opportunity of giving expression to their true nature, many a man has lived with laying hens for most of his life and several of theirs and never noticed what should have been most obvious. *It is far more important that a fowl should have shelter from the rain or snow throughout the day, when the wind can ruffle and displace her plumage, exposing her skin, than at night when, as I have been interested to observe on so many occasions, she can sit still, arranging all her feathers carefully and pointing her beak, for all the world as though in an act of worship, towards the heavens.*

THE END

This view of the interior of the Thompson modified henyard throws into relief the unique roosting arrangements. The inmates can please themselves whether they perch beneath a waterproof and snowproof roof or exposed to the heavens.

The bottomless nesting boxes are so designed that they can stand in full light without any of the eggs being visible from the outside.

In this view of the interior of an incubator house at Shipton Poultry Farm it can be seen how the Ironclads are arranged. Observe the size and height of the trolley at the far end. It runs on piano castors and a thick layer of heat-retaining felt rests on top in between a pair of wide boards which are kept waxed. Each egg tray is slid on to it for turning or testing; and it is much easier to keep moving it to where it is wanted than to have to hump loads of eggs repeatedly to some distant fixed bench.

Photo by Andrew Nicholson

ACKNOWLEDGMENTS AND COMMENTS

One morning, a year or two back now, a tall, striking-looking young couple turned up at this address out of the blue and without having troubled to fix an appointment or even to find out whether I wanted to see them at all or not.

The male end of this contingent had a large, brand new folder tucked under his arm and announced that he was 'researching a book on domestic poultry keeping', and would be grateful for any help I could give or anything I could show him.

My son took him on a tour of the stock and equipment situated within walking distance of the house and buildings, as I was busy in the office. And the only comment felt to be worth while repeating to me afterwards had to do with the fact that there appeared to be no hens around the place sitting on eggs or bringing up chicks.

But before that this young man, whose name I cannot recall, looked blank when it was mentioned that the less eggs, within certain limits, my breeders laid, the better pleased I was. Only after much work on my part did the fact sink in that it was stamina and quality rather than quantity that I was after.

My visitor then intimated, quite unnecessarily, that he knew nothing about poultry, which was why, knowing of my reputation, he had come to me. It would appear that it had occurred to him—or his publisher—that here was a market from which some useful pickings might be obtained. And the little book that eventually came into being bears every sign of having been got together by the expedient of picking my brains and those of one or two others here and there, and filling in the gaps from books, articles and suchlike.

I told the author that, if I knew nothing about a particular subject, I would not attempt to write about it, but would leave it to those who did. However I would tell him anything I could. Whatever he might succeed in turning out was hardly likely to possess enough scope to compete with or pose any serious threat to the work I was myself planning.

The pair finally departed, after promising faithfully to keep in touch and let me know how things turned out and what they did with the material; but that was the last I ever heard from them. When the little book appeared no attempt was made to acknowledge my help. And in fact the nearest the author came to rendering thanks consisted of a brief and grudging sentence in which I was accorded the role of 'a small supplier of day old chicks'. It was not thought worth while to mention the Dorset or to include my name in the list of those from whom suitable stock might be obtained.

I mention this since I happen to be a believer in giving credit and discredit where each is due, and because if there is a sense of indebtedness to anyone this should be recorded. As indicated in the text, most of what I have done, all the way through, that was of any consequence, has been carried out in spite of and not because of all the counsel, advice and exhortation that was offered or available. But, having said that, whatever debt I may owe to my grandfather and the late H. R. Hunter, to Henry and Ernest Robinson as well as the various other old-fashioned or new-fashioned poultrymen and general farmers, along with their wives and daughters, whom I was privileged to know back in the days when farming was farming and the wholesale retreat from hard physical work had not yet begun, I freely and gratefully acknowledge.

I also, whilst I am at it, desire to thank all the men and women from whose mistakes I have benefited, or who have provided valuable object lessons in what not to do and how not to look after poultry.

Nor must we leave out the handful of publishers who, when approached about a book of the kind you have just finished plodding through, gave me the cold shoulder. The debt which I owe to these last is not only considerable but in the nature of things one that no possibility exists of discharging. The reason for this deep sense of obligation is that their action and attitude has removed any temptation on my part to allow editorial or indeed any other kind of control out of my hands. To have done so, if not exactly a recipe for disaster, would at least have meant that I should have had to try to push somebody, almost certainly at a distance, around in an almost certainly futile attempt to get that done, and done in time, that I am perfectly capable of doing myself. Had I been injudicious enough to hand the text and the art work—that is, the whole bag of tricks—over to an outside firm the betting is that, instead of my telling them what I wanted, they would have ended up by trying to tell me what they wanted. Either that or they would have decided to appoint an organic/poultry editor—for which post, the form and content partaking of the nature that they do, I should have been the handiest and most obvious condidate. All of which brings us back with a bump to the familiar coals-to-Newcastle syndrome.

As it is I have the good fortune to live almost within walking distance of one of the best printers in the country, who began life as a farmer and could still, if put to it, like me and my bank manager, sit down and milk a whole row of cows by hand. What one doesn't know the other does; so between us we shall, I have every confidence, finish up with a very reasonable job.

Several of the leading publishers in this country have, indeed, over the years tried hard to get me to accept a commission to provide them with a textbook or handbook on poultry husbandry; but I refused to consider this. Textbooks are nothing in my line, quite aside from the fact that they would never have agreed to handle the only sort of book I would have been willing to write. Nor will I undertake a 'socially acceptable' work on any subject or one in which I am writing for a market and not because this is something that *ought* to be done, or where all or any of the punches are pulled and my tongue is stuck firmly in my cheek.

Many years ago a close neighbour of ours, having for reasons or from causes that were never vouchsafed, at least in my young and innocent presence, ended up with no sons and just one daughter, felt the trouble and expense of sending her to college to be worth while if she was going to inherit the farm one day. Soon after her return, according to the version retailed by my father, he came upon her in the dairy making a complete hash of some simple task that she should have been able to accomplish with ease long before she left home.

"Thou gurt, eddicated fule!" he exclaimed wrathfully.

A chap once wrote that, the more competent a publisher showed himself to be at publishing, the more useless and ham-handed he always appeared to be when it came to such basic and all-important tasks as sweeping the floor or making a cup of tea worth drinking. I am sure there is a moral hidden somewhere in this picture, if only one had time to stop and ferret it out. But what I have noticed, in a lifetime during which I have worked my way through many thousands of books, is that the more lofty the writer's perch amid the academic or scientific echelons, the less good he seems to be at arranging his work, dotting the i's, crossing the t's and so forth.

At least every other highbrow, intent upon distributing his cerebral offspring as far over the globe as possible, seems to have to have somebody at one elbow the whole time making suggestions, someone else at the other busily arranging all the raw material he has spewed forth, and his wife plus at least one other woman around: one putting his spelling right and the other doing the typing and generally licking the manuscript into the sort of shape the publisher and printer can make some sense of.

It is years since I last employed a secretary. I have to take care of my own spelling, cross my own t's, dot my own i's and do all my own typing. The other members of my family make each his or her own contribution. My wife, for instance, was responsible for the front cover design; and my son executed the line drawings.

I have talked about the various ones to whom I feel I owe a debt of one sort or another. Before closing the book there are four people whom I must pick out for special mention. These are: Robert Waller; Ailsa Pain; David Stickland and Dick Robinson. Without exception these genuine friends came into and remain in the organic movement from the best and highest of motives. And, time and again, they have provided the flint and steel required to strike just the right sort of sparks.

Colin Shaw, who has allowed me to dedicate this book to him, is one whose friendship I have always greatly treasured. Nowadays involved in administrative work as Director of Television, during his early days with the B.B.C. he produced the series of dramatised documentaries which I wrote about farming and country life in the North Riding of Yorkshire. It seems therefore fitting that, his name having been the first to appear in these pages, it should also be the last.

Matthew A. Thompson,

Tuesday, 14th March, 1978.   Shipton Poultry Farm.

SOME USEFUL ADDRESSES

The Henry Doubleday Research Association, 20 Convent Lane, Bocking, near Braintree, Essex.

The Soil Association, Walnut Tree Manor, Haughley, near Stowmarket, Suffolk.

FREGG (Free Range Egg Approval Scheme), 39 Maresfield Gardens, London, N.W.3.

Organic Farmers and Growers, Ltd., Martello House, 5 Station Road, Stowmarket, Suffolk.

The Farm and Food Society, 4 Willifield Way, London, N.W.II.